AF700705

Copyright

Copyright © 2025 David A. Fulcher. All rights reserved.

No part of this book may be copied, reproduced, distributed, or transmitted in any form or by any means — including photocopying, recording, or other electronic or mechanical methods — without the prior written permission of the publisher, except in the case of brief quotations embodied in critical reviews and certain other non-commercial uses permitted by copyright law.

This book is licensed for your personal use only. It may not be resold or given away to others. If you wish to share this book with someone else, please purchase an additional copy. If you're reading this book and did not purchase it, or it was not purchased for your use only, please return to an authorized retailer and purchase your own copy. Thank you for respecting the hard work of this author.

ISBN (Paperback): 978-1-7641531-5-7

For permissions or inquiries, please contact: David A. Fulcher;
dfulcher@immunology4bpts.com

Please Respect the Creative Process!

Thank you for purchasing 'MCQs in Clinical Immunobiology – Book 2'. I very much hope that it will be very valuable to your learning.

To those readers who have not obtained this book via legitimate avenues, though, I want you to be aware that creating educational materials such as this book takes **many hours of personal effort**, and I would really appreciate it if you acknowledged this by supporting this book financially. So:

If you have obtained this book illegally, please purchase a legal copy...

I cannot continue to generate teaching materials if they are copied the moment they are produced. If you think my teaching is valuable and would like to see it continue into the future, I ask you sincerely to support my work.

Please consider this request carefully before reading on.

Thankyou!

David Fulcher

Table of Contents

COPYRIGHT **1**

PLEASE RESPECT THE CREATIVE PROCESS! 2

FOREWORD **5**

FROM THE AUTHOR 5
ABOUT 'MCQS IN CLINICAL IMMUNOBIOLOGY' 6
RESOURCES 7
Primer of Clinical Immunobiology *8*
Immunology4physicians Video lecture series *8*
Immunology4physicians on Patreon *8*
Immunology4physicians on YouTube *9*
Instagram imm4physicians *9*

QUESTION PAPER **10**

WORKED ANSWERS **21**

QUESTION 1. PLASMA CELLS AND SERUM IGG MAINTENANCE 21
QUESTION 2. COMPLEMENT CONSUMPTION AND LOW C3/C4 LEVELS 22
QUESTION 3. COLD URTICARIA AND PHYSICAL URTICARIAS 23
QUESTION 4. AUTOIMMUNITY IN APS1/APECED 24
QUESTION 5. REGULATORY T CELLS AND FOXP3 26
QUESTION 6. NLRP3 IN GOUT 27
QUESTION 7. IMMUNOLOGICAL BASIS OF HYPER-IGE SYNDROME AND TH17 CELL DEFICIENCY 28
QUESTION 8. TARGETING T CELL ACTIVATION: THE ROLE OF ABATACEPT 30
QUESTION 9. RITUXIMAB EFFECTS IN AUTOIMMUNE DISEASE 32
QUESTION 10. IVIG USE IN SPECIFIC ANTIBODY DEFICIENCY 33
QUESTION 11. NK CELL ACTIVATION AND MISSING SELF 34
QUESTION 12. TAP DEFICIENCY AND CD8 T CELL DEVELOPMENT 35
QUESTION 13. T CELL-INDEPENDENT ANTIBODY PRODUCTION 37
QUESTION 14. CRYOGLOBULINEMIA TYPES AND RHEUMATOID FACTOR 38
QUESTION 15. PLASMACYTOID DENDRITIC CELLS AND VIRAL INTERFERON RESPONSE 39
QUESTION 16. HEREDITARY ANGIOEDEMA MANAGEMENT 40
QUESTION 17. CLASSIC MENDELIAN INHERITANCE AND IMMUNODEFICIENCIES 41
QUESTION 18. COMPLEMENT FACTOR C2 DEFICIENCY 43
QUESTION 19. PAROXYSMAL NOCTURNAL HEMOGLOBINURIA (PNH) 44
QUESTION 20. ACE INHIBITOR-INDUCED ANGIOEDEMA 45
QUESTION 21. MONOCLONAL ANTIBODIES ISOTYPE AND FUNCTION 46
QUESTION 22. CHRONIC MUCOCUTANEOUS CANDIDIASIS AND TH17 DEFECTS 47
QUESTION 23. PATHOGENIC VS MARKER AUTOANTIBODIES 48
QUESTION 24. IGA DEFICIENCY 49
QUESTION 25. V(D)J RECOMBINATION AND RAG PROTEINS 50
QUESTION 26. SARS-COV-2 AND THE INFLAMMASOME 51
QUESTION 27. DISTINGUISHING CVID FROM SECONDARY CAUSES 52
QUESTION 28. TARGETS OF B-CELL DEPLETING MONOCLONALS 53
QUESTION 29. INDICATIONS FOR B-CELL DEPLETING MONOCLONALS 54
QUESTION 30. TEZEPELUMAB AND ALLERGIC DISEASE 56
QUESTION 31. TELITACICEPT AND B-CELL MATURATION 57

QUESTION 32. JAK INHIBITORS AND GIANT CELL ARTERITIS 59
QUESTION 33. ALLERGIES AND PERSISTENCE WITH AGE 60
QUESTION 34. T CELLS AND ORAL TOLERANCE 61
QUESTION 35. COMPLEMENT INHIBITORS AND IGA NEPHROPATHY 62
QUESTION 36. BISPECIFIC T CELL ENGAGERS: BLINATUMOMAB 64
QUESTION 37. NK CELL DEFICIENCY AND HERPES INFECTIONS 65
QUESTION 38. THERAPEUTICS TARGETING CYTOKINES 66
QUESTION 39. TREX1 DEFICIENCY AND CYTOPLASMIC DNA 68
QUESTION 40. T CELL ACTIVATION 70
QUESTION 41. T-FOLLICULAR HELPER CELLS AND B CELL DIFFERENTIATION 71
QUESTION 42. MASTOCYTOSIS AND DIAGNOSTIC TESTING 72
QUESTION 43. T CELL PRECURSORS IN THE THYMUS 73
QUESTION 44. CVID VS X-LINKED AGAMMAGLOBULINEMIA 74
QUESTION 45. IMMUNE THROMBOCYTOPENIA (ITP) IN CVID 75
QUESTION 46. TREC NEWBORN SCREENING FOR SCID 76
QUESTION 47. ANTIBIOTIC ALLERGY AND CEPHALOSPORIN CROSS-REACTIVITY 78
QUESTION 48. UNC93B1 MUTATIONS AND LUPUS PATHOGENESIS 79
QUESTION 49. SCID MUTATIONS AND KEY MOLECULAR DEFECTS 80
QUESTION 50. DEPEMOKIMAB AND EOSINOPHILIC ASTHMA 82
FINAL REMARKS 83

FOREWORD

From the author

My medical degree was from the University of Sydney, graduating in 1984. I then specialised in Clinical Immunology (Fellowship of the Royal Australasian College of Physicians) and Immunopathology (Fellowship of the Royal College of Pathologists of Australasia) and went on to do a PhD in B-cell tolerance at what was to become the Centenary Institute, Sydney (1996).

My clinical work has extended over 30 years, including specialist practice in clinical and laboratory immunology, and within public and private settings. Currently I am lead Immunopathologist at 4Cyte Pathology.

I have had a **lifetime passion for teaching Immunology**, applying the basic biology of the immune system to understanding immune disease. This has involved me in many teaching roles over the last three decades, the most important of which was a weekend 'crash course' in Immunology for doctors specialising in internal medicine, called "**immunology4bpts**", which combined teaching of the basics of immunobiology with clinical immunology. This grew in popularity, expanding to over 500 registrants, and run over two weekends, one in Sydney and the other in Melbourne; since COVID, it moved on-line as video-on-demand. The **immunology4bpts** course has now been running for over 20 years.

I have worked at Australian National University as Professor (2016-2021), then Honorary Professor (2021-2025), where I rejuvenated the 'Advanced and Applied Immunology' course which I convened for five years (2016-2021). At ANU I was also responsible for the teaching outreach arm of the 'Centre for Personalised Immunology' and supervised the Flow Cytometry data as a Chief Investigator in the Centre for Research Excellence NHMRC grant. Prior to that, I was Associate Professor at University of Sydney for ~10 years.

I have held every major Australian teaching position in Immunology and Immunopathology, including:

- Co-ordinator of Advanced Training in Immunology;
- Chair of the Joint Specialist Advisory Committee in Immunology (RACP);
- Chief Examiner in Immunopathology (RCPA).

Teaching in Internal Medicine has also included roles as:

- Director of Physician Training (Western Sydney)
- Network Director of Physician Training (Western Sydney)

Dr David A. Fulcher. Clinical Immunologist and Immunopathologist
MBBS (Hons II), PhD, FRACP, FRCPA.

About 'MCQs in Clinical Immunobiology'

The principle behind my teaching of Immunology to doctors is based on my philosophy that immunobiology becomes much more interesting to doctors if the fundamentals of immunobiology are presented in the context of their clinical implications. Let me explain...

Initially '**immunology4bpts**' weekend training course separated out the basic immunobiology from clinical medicine, but I noticed that there was an enormous sigh of relief when the basic biology bit was finished, and the medical audience could move into their comfort zone i.e. doctoring! So, I decided to completely intermix the two streams; the basic biology was immediately illustrated by the clinical medicine, and back again, so as to maintain engagement of the audience throughout the weekend. Thus was born **Clinical Immunobiology!**

The course was so successful that it led to my writing the '*The Primer of Clinical Immunobiology*' to support the lecture series:

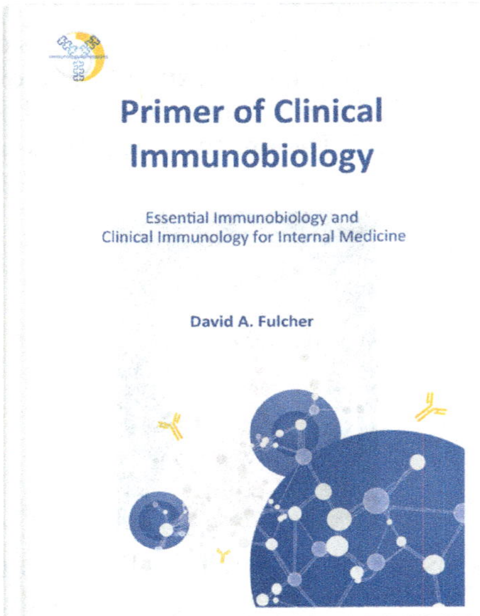

And because the course has always finished with a 50-question examination with worked solutions, I have written the *'MCQs in Clinical Immunobiology'* series as book equivalents to that quiz.

How to approach this book

Obviously the first step is to do the exam paper, which is in the next section. If you want to work against the clock, allow yourself about 30 minutes for the 50 questions, which is about 36 seconds per question! But if you want to do it in more palatable chunks, that is also fine by me!

In either case, commit your answers to paper, and when you are ready to check them, head to the 'Worked Answers' section. And with each answer, you will find a reference to one or more chapters of the '*Primer of Clinical Immunobiology*' which you can consult to remind yourself of the material.

Resources

These are the complementary learning resources that I manage, and how they fit together:

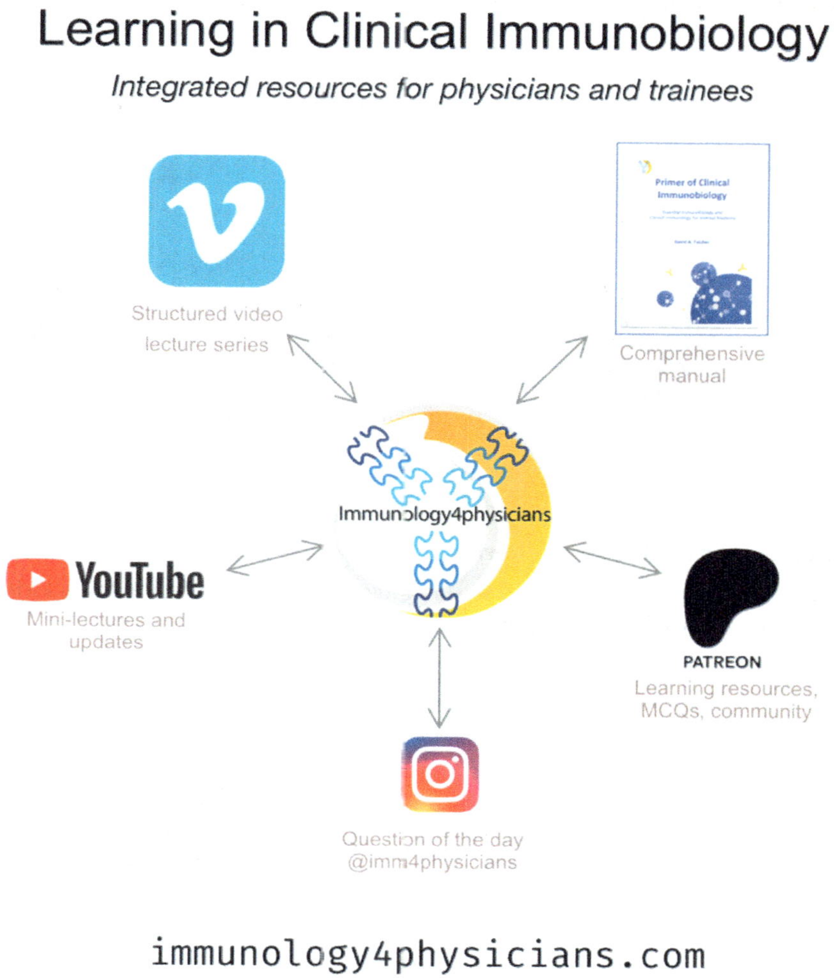

Primer of Clinical Immunobiology

This is the source text for learning Clinical Immunobiology, and you will find chapter references throughout this book. It covers what all internists and trainees in internal medicine need to know about basic and clinical immunology:

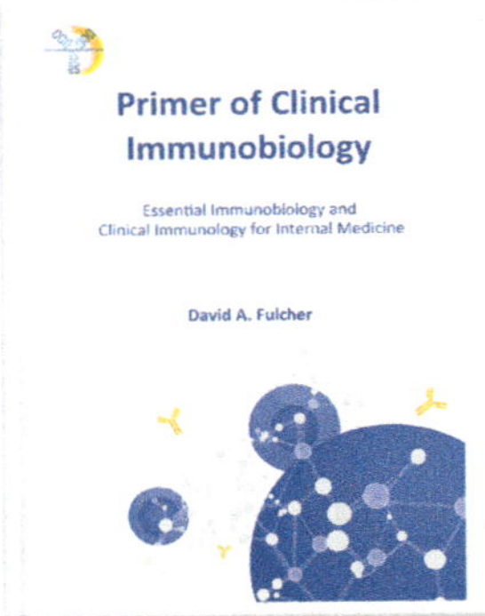

Print edition (paperback and hardback) — amazon.com (US store)

eBook: Apple books; Kindle (App)

All links at:

immunology4physicians.com

Immunology4physicians Video lecture series

Each chapter of the '*Primer of Clinical Immunobiology*' has a corresponding video lecture that you can rent-stream from Vimeo:

immunology4physicians.vhx.tv/products

Immunology4physicians on Patreon

This site forms the main teaching hub for 'Clinical Immunobiology' and consists of a series of learning exercises to reinforce teaching materials in the *Primer of Clinical Immunobiology* and the video lecture series.

patreon.com/immunology4physicians

The site includes a wide range of learning activities, including:

- Revision exercises. These make use of news or recent publications to illustrate underlying fundamental principles of Clinical Immunobiology.
- Mini-lectures. These are short presentations designed to explain one specific topic in clinical immunobiology.
- Case presentations
- Multiple Choice Questions: groups of five MCQs with worked answers.

Standard membership is free, allowing access to a few of the activities, but a small monthly subscription unlocks access to the entire catalogue of over 100 learning exercises over more than two years. All the profits go to the World Wildlife Fund—Australia.

Immunology4physicians on YouTube

You will find here short revision lectures (10-15 mins) of various topics in Clinical Immunobiology that I could not fit into the course!

youtube.com/@immunology4physicians

Instagram imm4physicians

Follow me on Instagram for 'Clinical Immunobiology Question of the Day':

@imm4physicians

instagram.com/imm4physicians

And with those resources behind you, it is now time to start the Quiz!

QUESTION PAPER

1 (DAF*0172:2008). Serum IgG levels are maintained by the secretion of immunoglobulins by

A. activation of newly-generated naive B cells
B. re-stimulation of memory B cells
C. generation of short-lived plasmablasts in the spleen
D. persistence of long-lived plasma cells in the marrow

2 (DAF*0174:2008). Decreased levels of C3 and C4 are typical of all the following except:

A. SLE
B. Serum sickness
C. Chronic active hepatitis B infection
D. Cryoglobulinaemia

3 (DAF*0176:2008). A 30 year old woman presents to ED after collapsing after swimming in a river. During subsequent out-patients assessment, an ice-cube is placed on the forearm, and a large wheal develops at the site. What is the most likely diagnosis?

A. Chronic spontaneous urticaria / angioedema
B. Cold urticaria
C. Cryoglobulinaemia
D. Cold agglutinin disease

4 (DAF*0178:2008). Autoimmunity in 'autoimmune polyendocrinopathy type I' (APECED) is due to:

A. a failure of deletion of autoreactive B cells in the bone marrow
B. a failure of negative selection in the thymus
C. molecular mimicry between a foreign antigen and a self antigen expressed widely in endocrine tissues
D. dysfunctional regulatory T cells

5 (DAF*0184:2009). The most specific marker of a natural (central) regulatory T cell is:

A. CD4
B. CD25
C. CTLA4
D. FoxP3

6 (DAF*0187:2009). In the acute arthritis accompanying gout, the link between urate deposition and inflammation is provided by which immunoreceptor family of the innate immune system?

A. NOD-like receptor (NLRs)
B. Toll-like receptor (TLRs)
C. Macrophage-mannose receptor (MMR)
D. RIG-like receptor

7 (DAF*0189:2009). Hyper-IgE syndrome results from a heterozygous loss-of-function mutation in Stat3, an intracellular signalling molecule distal to receptors for IL-6, IL-1 and IL-23. Predisposition to Candida and Staphylococcal infections results from deficiency of which of the following T-cell subsets?

A. Th1
B. Th2
C. Th17
D. Tfh

8 (DAF*0192:2009). Which of the following potential treatments for lupus is aimed specifically at inhibiting T-cell activation?

A. Rituximab
B. Anti-BAFF (/BlyS)
C. Abatacept (CTLA4-Ig)
D. Anti-PD-1

9 (DAF*0194:2009). Following the successful use of Rituximab (anti-CD20) for pemphigus vulgaris, an autoimmune disease of the skin, all of the following changes would be expected EXCEPT:

A. Decrease in titres of autoantibodies
B. Decrease in tetanus-specific antibodies
C. Decrease in CD19+ cells (% lymphocytes)
D. Increase in CD3+ cells (% lymphocytes)

10 (DAF*0196:2009). Intravenous immunoglobulin is indicated in Australia for patients with a normal total IgG level, but who nevertheless have humoral immunodeficiency as evidenced by:

A. IgA deficiency
B. deficiency of both IgA and one or more IgG subclasses
C. Poor pneumococcal vaccine responses
D. low switched memory B cells

11 (DAF*0197:2009). NK cell killing of a target cell is most likely if the target cell:

A. expresses peptide antigen presented on HLA class II
B. expresses peptide antigen presented on HLA class I
C. down-modulates HLA class I
D. down-modulates HLA class II

12 (DAF*0201:2009). Deficiencies of Transporter-associated proteins TAP-1 and TAP-2 result in a deficiency of CD8 T cells, because they are essential for:

A. expression of HLA class I on thymic epithelial cells in primary lymphoid tissue
B. expression of HLA class II on thymic epithelial cells in primary lymphoid tissue
C. expression of HLA class I by antigen-presenting cells in secondary lymphoid tissue
D. CD8 T cell exodus from the thymus

13 (DAF*0203:2009). Production of which of the following immunoglobulin isotypes is NOT T-cell dependent?

A. IgE
B. IgM
C. IgA
D. IgG

14 (DAF*0206:2009). A patient with a peripheral vasculitic rash is found to have cryoglobulinaemia. The cryogloublin is found to consist of a monoclonal IgM paraprotein and polyclonal IgG. Which is the following tests is most likely to be positive?

A. RF
B. SSA
C. dsDNA
D. HBV-DNA

15 (DAF*0209:2009). Which of the following cell types is most responsible for the interferon response to viral infections?

A. Monocytes
B. Macrophages
C. Plasmacytoid dendritic cells
D. B cells

16 (DAF*0211:2009). A 26 year old woman presents to casualty for the fourth time in two years with spontaneous development of lip swelling. Labs have previously shown absent serum C4, and reduced levels of C1 inhibitor and function. Which is the most appropriate long-term management:

A. Intermittent steroids for acute attacks
B. Weekly infusions of C1 inhibitor concentrate
C. Fortnightly injections of lanadelumab
D. Icatibant to be taken at the earliest sign of severe swelling

17 (DAF*0213:2009). Classic Mendelian inheritance is characteristic of all the following immunodeficiencies EXCEPT:

A. Common Variable Immunodeficiency
B. Hyper IgM syndrome
C. Chronic granulomatous disease
D. Bruton's agammaglobulinaemia

18 (DAF*0215:2009). A deficiency of C2 is most likely to result in:

A. Recurrent attacks of angioedema
B. Neisseria infections
C. Mesangiocapillary glomerulonephritis type II (dense deposit disease)
D. Lupus-like disease

19 (DAF*0216:2009). Paroxysmal Nocturnal Haemaglobinuria occurs as a result of

A. A genetic lack of transmembrane-bound complement control proteins
B. An acquired lack of transmembrane-bound complement control proteins
C. A genetic lack of GPI-linked complement control proteins
D. An acquired lack of GPI-linked complement control proteins

20. (DAF*0217:2025). A 72 year old man presents to emergency with acute swelling of the tongue. He has a history of hypertension, ischaemic heart disease, and type 2 diabetes. Which of the following is the most likely cause for the tongue swelling?

A. Amyloidosis
B. C1 inhibitor deficiency (hereditary angioedema)
C. Adverse drug reaction to perindopril
D. Idiopathic (spontaneous) angioedema

21 (DAF*0226:2010). A monoclonal antibody with the LEAST chance of causing immune activation will be of which isotype?

A. IgG1
B. IgG2
C. IgG3
D. IgG4

22. This photo shows candida-related onycholysis of the toenails. Inherited deficiency of which protein could explain this condition in a 16 year old female?

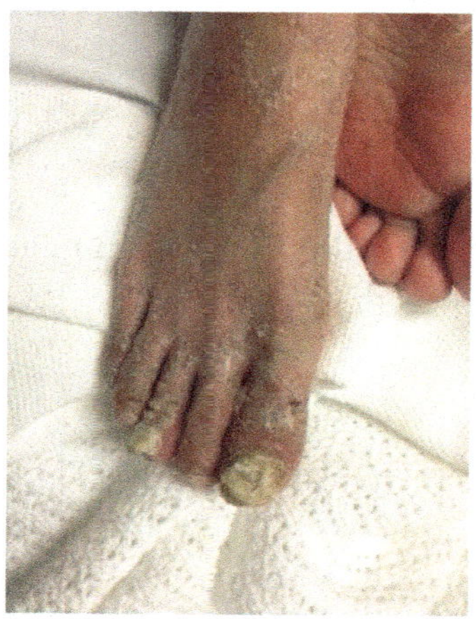

A. Mannose-binding lectin (MBL)
B. Gamma-common chain
C. CD40L (CD154)
D. IL-17 receptor

23 (DAF*0236:2010). An autoantibody is shown to be pathogenic if all the following apply EXCEPT:

A. It causes disease in neonates of affected mothers
B. It is positive in almost all patients with an autoimmune disease
C. Levels fluctuate with disease activity
D. It causes disease on injection into mice

24 (DAF*0238:2010). Which of the following statements about IgA deficiency is INcorrect?

A. It is typically asymptomatic
B. It may be genetically associated with common variable immunodeficiency
C. It may present as a transfusion reaction
D. It may result in false positive coeliac serology

25 (DAF*0243:2010). Binding of Recombinase Activating Gene (RAG) proteins to V, D and J sequences is critical for the development of:

A. just T cells
B. just B cells
C. T & B cells
D. all lymphocytes

26 (DAF*0919:2025). Viroporin 3a is a protein of the SARS-CoV-2 virus which can activate NLRP3. The effect of this activation would be release of:

A. IL-2
B. IL-4
C. IL-1
D. TGF-beta

27 (DAF*0920:2025). Which of the following would negate a diagnosis of Common Variable Immunodeficiency (CVID)?

A. Normal IgA
B. Normal memory B cells
C. Thymoma
D. Normal dynamic vaccine responses

28 (DAF*0921:2025). Inebilizumab is a humanized IgG1κ monoclonal antibody against CD19 that is effective in the treatment of myasthenia gravis. Which of the following B-lineage cells would be targeted by inebilizumab but unaffected by rituximab?

A. Mature B cells
B. Immature B cells
C. Memory B cells
D. Pro-B cells

29 (DAF*0922:2025). B-cell depleting monoclonal antibodies have been shown to be effective in all the following EXCEPT:

A. Myasthenia Gravis
B. IgG4 disease
C. Ankylosing spondylitis
D. Multiple sclerosis

30 (DAF*0923:2025). Tezepelumab is an anti-TSLP monoclonal antibody which is effective in the treatment of severe chronic rhinosinusitis with nasal polyposis and asthma. This therapeutic acts at the level of:

A. Innate Lymphoid Cells (ILCs)
B. Th1 cells
C. Th2 cells
D. Cytotoxic T cells

31 (DAF*0924:2025). Telitacicept is a fusion protein of TACI linked to Fc end of IgG which has been trialled in lupus. Which aspect of B-cell biology does this inhibitor target:

A. B-cell generation in the marrow
B. The passage of B-cells through the transitional stage
C. B cell activation by T cells
D. Establishment of antibody forming cells in bone marrow niches

32 (DAF*0925:2025). Upadacitinib is of potential benefit in the management of patients with giant cell arteritis, mostly likely because of its effect on:

A. Blocking IL-6
B. Blocking the IL-6 receptor
C. Blocking the downstream signalling of the IL-6 receptor
D. Blocking release of IL-6 from macrophages

33 (DAF*0926:2025). Hypersensitivity reactions to which of the following allergens is LEAST likely to decline with age?

A. Penicillin
B. Peanut
C. Milk
D. Soy

34 (DAF*0927:2025). T-cells play a pivotal role in maintenance of tolerance to food antigens. The T-cell subset most associated with this role produces which of the following cytokines?

A. Interferon-alpha
B. Interleukin-22
C. Interleukin-10
D. Interleukin-gamma

35 (DAF*0928:2025). Iptacopan is an oral inhibitor of Complement Factor B that has beneficial effects in IgA nephropathy. This is consistent with the observation that IgA nephropathy is a disorder driven by:

A. the lectin pathway of complement activation
B. the classical pathway of complement activation
C. the alternate pathway of complement activation
D. the membrane attack complex

36 (DAF*0929:2025). Blinatumomab is a bispecific, single-chain molecule that targets CD19 and CD3 ("BITE"). Such a therapeutic could be of potential therapeutic utility in which of the following conditions?

A. T-cell lymphoblastic lymphoma
B. NK-LGL leukaemia
C. Acute myeloid leukaemia
D. B-cell acute lymphoblastic leukaemia

37 (DAF*0930:2025). NK cell deficiency should be suspected in any patient with early onset:

A. Herpes virus infections
B. Pneumocystis pneumonia
C. sinopulmonary infections
D. Neisseria infections

38 (DAF*0931:2025). Which of the following therapeutics targets only a single cytokine or cytokine action?

A. Upadacitinib (Jakinib)
B. Dupilumab (anti-IL4R-alpha)
C. Ustekinumab (anti-IL-12p40)
D. Benralizumab (anti-IL5R-alpha)

39 (DAF*0932:2025). TREX1 deficiency is a rare cause of an early-onset lupus-like disorder. Its deficiency allows accumulation of DNA in the cytoplasm, which likely contributes to lupus pathogenesis by:

A. activating anti-DNA-specific B-cells on cell lysis
B. binding to innate nucleotide pattern-recognition receptors
C. binding to toll-like receptors
D. stimulating formation and triggering of NLRs in inflammasomes

40 (DAF*0933:2025). In terms of T-cell activation, which of the following surface molecules is the 'odd one out'?

A. LAG-3
B. CTLA4
C. CD28
D. PD-1

41 (DAF*0934:2025). Which of the following functions is NOT characteristic of T-follicular helper cells?

A. B-cell differentiation to short-lived plasma cells
B. Production of high affinity antibodies by B cells
C. Class-switch recombination in B cells
D. Differentiation of naive to memory B cells

42 (DAF*0935:2025). An 18-year-old male with a long-standing history of presumed food allergies and recurrent anaphylaxis was referred for evaluation. Since infancy, he had experienced wheezing, rhinitis, and skin lesions, with episodes of flushing, angioedema, and presyncope often attributed to food triggers. Despite extensive dietary restrictions and treatments, including omalizumab, his symptoms persisted. Physical examination revealed widespread keratosis pilaris and nodular skin lesions exhibiting Darier's sign. Serum basal tryptase was 23 ng/mL (NR<11).

Which of the following would be least likely to confirm the clinical diagnosis?

A. Skin biopsy
B. Bone marrow biopsy for KIT D816V mutation
C. 24hour urinary histamine
D. Serum chromogranin A level

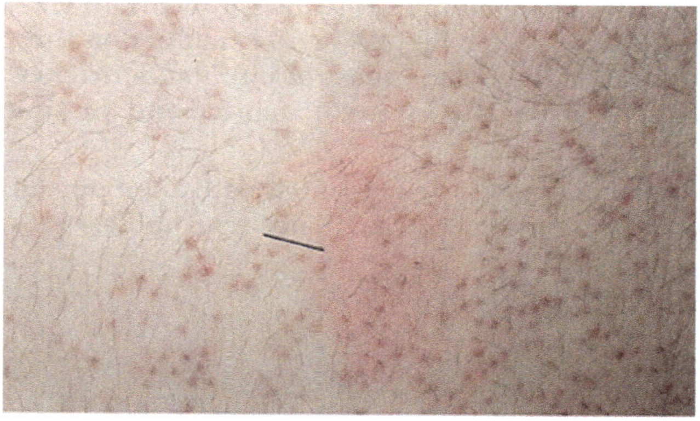

[Image from Azenha, R. T., Martins, D., Gomes, N., Pinheiro, J., Nogueira, A., Delgado, L., Plácido, J. L., & Coimbra, A. (2021). Case Report: Mastocytosis: The Long Road to Diagnosis. Frontiers in Immunology, 12, Article 635909. https://doi.org/10.3389/fimmu.2021.635909. Licensed under Creative Commons Attribution 4.0 International (CC BY 4.0). Image has been cropped from original but not otherwise altered]

43 (DAF*0936:2025). The earliest T-cell progenitor cells entering the thymus express:

A. both CD4 and CD8
B. CD4 only
C. CD8 only
D. neither CD4 nor CD8

44 (DAF*0937:2025). A 40-year-old male with a history of recurrent respiratory infections since age 5 was diagnosed with common variable immunodeficiency (CVID) at age 23, when he presented with severe obstructive lung disease and markedly low immunoglobulin levels (IgG <0.2 (NR>7g/L), IgA <0.2 (NR>0.7g/L), and IgM 0.22 (NR>1.4 g/L)).

Despite immunoglobulin replacement therapy, which reduced infection frequency, he continued to suffer from pulmonary insufficiency. Physical exam showed pulmonary rhonchi, and chest imaging revealed hyperexpanded lungs. Pulmonary function was severely impaired (FEV1 at 24.3% predicted). Flow cytometry of lymphocytes revealed 77% CD3 cells, CD19 0%, and NK cells 20.3%. Family history was negative for immunodeficiency.

Which of the following tests would be most relevant to confirm or refute CVID as the diagnosis?

A. Tetanus toxoid antibody response after vaccination
B. Memory B cell studies
C. CT chest
D. Btk gene sequencing

45 (DAF*0938:2025). A 65 year old woman was referred following the discovery of pan-hypogammaglobulinaemia on investigation for recurrent sinus and chest infections (IgG 0.6 g/l; IgA 0.16 g/l; IgM 0.44 g/l). Examination showed low-grade generalised lymphadenopathy. Other causes of immunodeficiency were excluded and a diagnosis of CVID made. IVIG was commenced and the sinus and chest symptoms improved.

Eleven months after the diagnosis she presented with a flat purpuric rash affecting her legs and blisters in her mouth. The next most relevant test is:

A. skin autoantibodies
B. measurement of switched memory B cells
C. platelet count
D. mouth biopsy

46 (DAF*0939:2025). The frequency of TRECs (T-cell receptor excision circles) in newborn dried blood samples is emerging as a screening test for early diagnosis of SCID. The test itself is based on which physiological T-cell developmental process?

A. Negative selection
B. Formation of a T-cell receptor
C. Somatic hypermutation
D. Alpha-beta versus gamma-delta TCR decision

47 (DAF*0940:2025). A 24 year old man with cystic fibrosis presents with respiratory infection for antibiotic treatment. He gives a past history of pencillin allergy following a systemic anaphylaxis after pencillin G; skin tests were performed and were positive.

Which of the following would be the safest antibiotic choice?

A. ticarcillin
B. ceftriaxone
C. cephalothin
D. cephalexin

48 (DAF*0941:2025). During the past year, four studies have reported ten mutations in UNC93B1, which encodes the Toll-like receptor (TLR) chaperone protein UNC93B1. All variants increased TLR7 and TLR8 signalling and caused systemic lupus erythematosus in young individuals. Such heightened signalling may explain the common association between lupus and:

A. the type I interferon signature
B. reduced complement levels
C. high CRP
D. hypergammaglobulinaemia

49 (DAF*0942:2025). Absence of circulating T cells might result from mutations in all the following EXCEPT:

A. RAG-1
B. CD40 ligand
C. IL-7 receptor alpha chain
D. Jak3

50 (DAF*0943:2025). Depemokimab is a long-acting monoclonal antibody to IL-5. As such, it is a potential therapeutic in the treatment of which of the following conditions?

A. Psoriatic spondyloarthropathy
B. Crohn's disease
C. Rheumatoid arthritis
D. Eosinophilic asthma

WORKED ANSWERS

Question 1. Plasma Cells and Serum IgG Maintenance

(see Primer of Clinical Immunobiology «Chapter—How B cells are Activated»)

1 (DAF*0172:2008). Serum IgG levels are maintained by the secretion of immunoglobulins by

A. activation of newly-generated naive B cells
B. re-stimulation of memory B cells
C. generation of short-lived plasmablasts in the spleen
D. persistence of long-lived plasma cells in the marrow

Answer D

When a B cell is activated, it undergoes isotype switching, somatic hypermutation, selection, and then ultimately differentiates into either a plasma cell or a memory B cell. Plasma cells migrate to the marrow, where they can continue producing specific immunoglobulins for years. So this is where the steady state IgG levels in plasma come from.

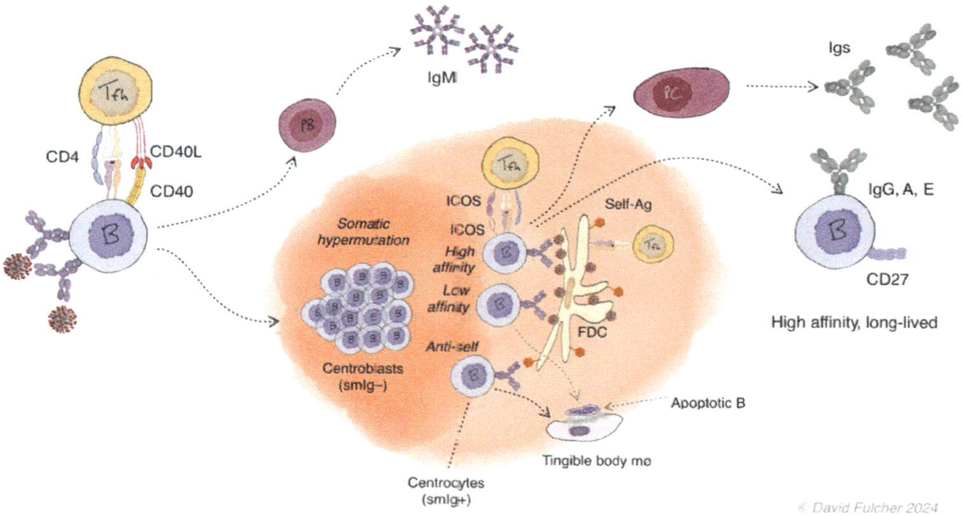

As for the other options in this question:

- Newly-generated naive B cells are not continually activating under resting conditions and are therefore not responsible for long-term IgG secretion.
- Memory B cells require re-exposure to antigen for activation. They may contribute to another wave of high affinity antibody responses, if reactivated by repeat infection, but do not contribute to the maintenance of serum IgG.

Note in passing: there is an alternative outcome for activated B cells, namely to differentiate to plasmablasts. Plasmablasts localise adjacent to the site of activation and rapidly produce IgM, typically without isotype switching or affinity maturation, and they are therefore responsible for that early antibody response.

Question 2. Complement Consumption and Low C3/C4 Levels

(see Primer of Clinical Immunobiology «Chapter—Complement Disorders»)

2 (DAF*0174:2008). Decreased levels of C3 and C4 are typical of all the following except:

A. SLE
B. Serum sickness
C. Chronic active hepatitis B infection
D. Cryoglobulinaemia

Answer C

Decreased levels of C3 and C4 are absolutely characteristic of a few important immune conditions, and you should probably be able to reel off at least the top three (maybe five...). Complement consumption, reflected by a drop in C3 and C4, is seen in lupus, serum sickness, and cryoglobulinaemia. It's not universal, as levels can vary with disease activity, especially in lupus, but it's a very typical feature.

In contrast, complement consumption is not typical of chronic active hepatitis B infection; if you chose this option, maybe you were getting it mixed up with hepatitis C infection (e.g. HCV-related cryoglobulinaemia)?

Have a look at the following table from the Primer, where you'll find all of the major causes laid out for you!

Conditions affecting C3 and C4 levels		
↑C3, C4	Pregnancy Chronic inflammation Infection	
↓C3, C4	Sample mishandling	(C' is heat labile)
	Consumption	Immune disorders • SLE, vasculitis, serum sickness • Glomerulonephritis: mesangiocapillary or post-infectious glomerulonephritis (↓C3>C4) • Cryoglobulinaemia (↓C4>C3) • C1 INH deficiency (C4 only) • Chronic bacterial or parasitic infection e.g. infective endocarditis • Septic shock Mechanical • Radiocontrast dyes • Haemodialysis • Cardiopulmonary bypass
	↓Production	C4 deficiency C3 deficiency Malnutrition Liver disease
	↑Loss	Nephrotic syndrome Burns Enteropathy

Question 3. Cold Urticaria and Physical Urticarias

(see Primer of Clinical Immunobiology «Chapter—Non-Atopic Allergic Disease»)

3 (DAF*0176:2008). A 30-year-old woman presents to ED after collapsing after swimming in a river. During subsequent outpatients assessment, an ice-cube is placed on the forearm, and a large wheal develops at the site. What is the most likely diagnosis?

A. Chronic spontaneous urticaria / angioedema
B. Cold urticaria
C. Cryoglobulinaemia
D. Cold agglutinin disease

Answer B

This scenario isn't specifically covered in the Primer, but you should be able to get to work it out! So to summarise the scenario here, we're dealing with an anaphylactic-type event after cold exposure, and then a large wheal and flare that develops in response to a cold stimulus; this combination is quite characteristic of cold urticaria.

Cold urticaria is one of the physical urticarias, in which a physical trigger sets off the patient's symptoms. The four main types of physical urticarias are cholinergic (triggered by sweating), solar (triggered by sun exposure), aquagenic (triggered by water exposure), and cold urticaria (triggered by cold exposure). Of these, cholinergic urticaria is probably the most common, although cold urticaria is not infrequent!

In cold urticaria the reaction can be life-threatening should the patient do something unfortunate like jump into cold water; such a strong trigger can result in massive mediator release and shock.

Of the other options:

- Chronic urticaria is not associated with cold as a trigger, although the symptoms can worsen with physical pressure, suggesting some physical component.
- Cryoglobulinemia doesn't present with collapse; think Raynaud's, small vessel vasculitis, glomerulonephritis, skin infarctions.
- Cold agglutinin disease is a haematological disorder; it's got "cold" in the name, but it's a haemolytic process caused by clumping of red cells in the cooler extremities. There can be skin changes but these are chronic, and again acute collapse is not part of the syndrome.

Question 4. Autoimmunity in APS1/APECED

(see Primer of Clinical Immunobiology «Chapter—T-cell Development»)

4 (DAF*0178:2008). Autoimmunity in 'autoimmune polyendocrinopathy type I' (APECED) is due to:

A. a failure of deletion of autoreactive B cells in the bone marrow
B. a failure of negative selection in the thymus
C. molecular mimicry between a foreign antigen and a self antigen expressed widely in endocrine tissues
D. dysfunctional regulatory T cells

Answer B

APECED (or APS-1) is an autosomal recessive deficiency of AIRE (autoimmune regulator) in which there is a failure of negative selection in the thymus. This happens because AIRE is responsible for expression of tissue-specific self-antigens within the thymus, and when absent, potentially autoreactive T cells do not 'see' these self-antigens during development, and fail to be deleted. They can then escape into the periphery where they can drive a range of autoimmune (mostly endocrine) diseases.

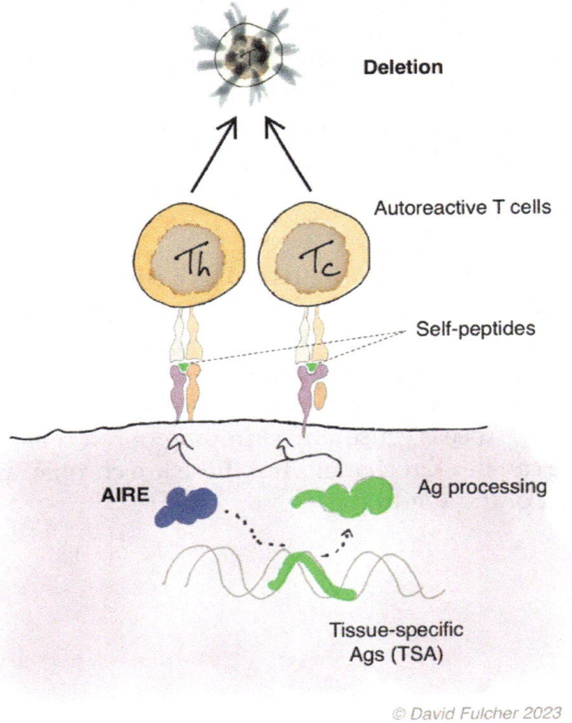

On the other hand, AIRE has nothing to do with B cells, Treg cells nor is it involved in the process of molecular mimicry.

The three most characteristic manifestations of APECED are hypoparathyroidism, Addison's disease, and chronic mucocutaneous candidiasis (CMC); CMC here is caused by autoantibodies against Th17-derived cytokines (IL-17 and 22), making it an example of an autoantibody-mediated immunodeficiency. Of course not all patients get all three manifestations; there are a lot of other autoimmune conditions that may be found, meaning that factors other than a failure of central tolerance are important.

Question 5. Regulatory T Cells and FOXP3

(see Primer of Clinical Immunobiology «Chapter—T-cell Tolerance»)

5 (DAF*0184:2009). The most specific marker of a natural (central) regulatory T cell is:

A. CD4
B. CD25
C. CTLA4
D. FoxP3

Answer D

This is a pretty good question, don't you think? (thank you!) because it helps revise the important surface markers on T cells in general:

- CD4 is certainly present on regulatory T cells, but is also expressed by all helper T cells, meaning it's not specific to Tregs.
- CD25, the α chain of the IL-2 receptor, can also be expressed on activated conventional T cells, not just regulatory T cells.
- CTLA-4 appears on late activated and exhausted T cells, so again, its expression is not specific to Tregs.

In contrast, FoxP3 is absolutely the most specific marker for regulatory T cells and it's FoxP3 that truly defines them. Other characteristics of Tregs are depicted in this diagram:

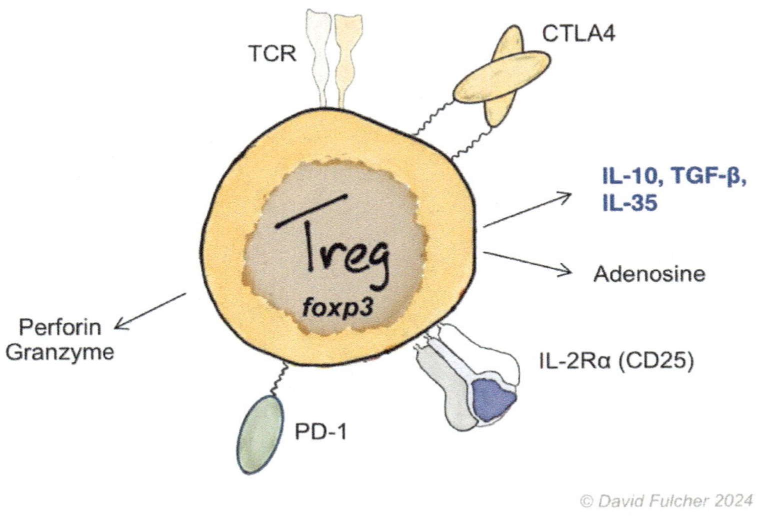

Question 6. NLRP3 in Gout

(see Primer of Clinical Immunobiology «Chapter—Innate Receptors»)

6 (DAF*0187:2009). In the acute arthritis accompanying gout, the link between urate deposition and inflammation is provided by which immunoreceptor family of the innate immune system?

A. NOD-like receptor (NLRs)
B. Toll-like receptor (TLRs)
C. Macrophage-mannose receptor (MMR)
D. RIG-like receptor

Answer A

You know all about innate immunoreceptors and how they act to detect motifs on foreign pathogens, but here is a classical example of how they can also be involved in the pathogenesis of disease. In gout, urate crystals form and are engulfed by macrophages. Due to their nature, the crystals break through the endosomal compartment and reach the cytoplasm. Meanwhile some secondary trigger induces the production of NLRP3 which assembles into an inflammasome. The inflammasome recognises the urate crystals as foreign and fires off, activating caspase-1, which cleaves pro-interleukin-1 into active interleukin-1β. IL-1β is a strongly pro-inflammatory cytokine that drives the inflammatory response:

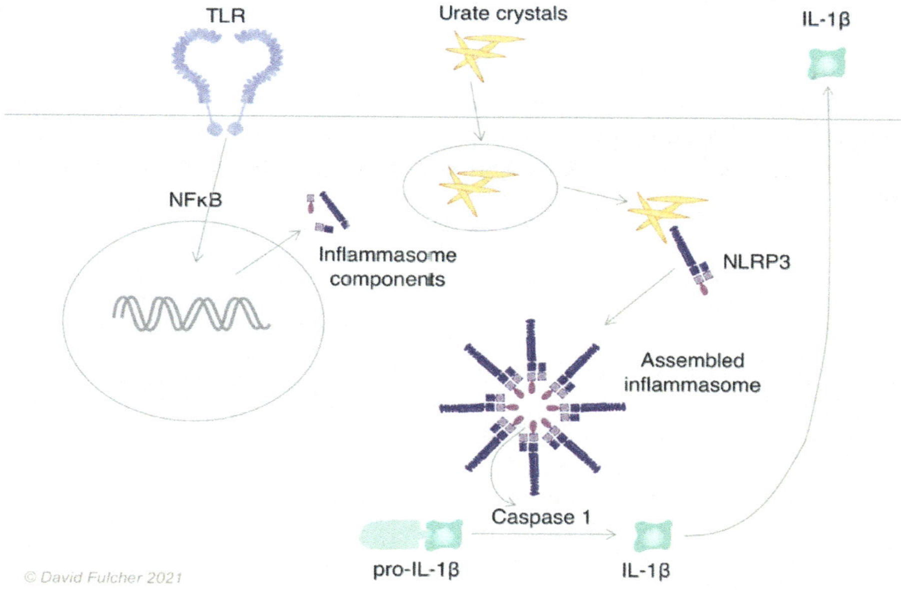

With regard to the other options:

- Toll-like receptors (TLRs) are also pro-inflammatory, but are not activated by urate crystals;
- Macrophage-mannose receptor (MMR) is a C-type lectin that binds pathogens but does not signal.
- RIG-like receptors recognise viral RNA, not urate crystals, therefore they play no known role in gout.

Question 7. Immunological Basis of Hyper-IgE Syndrome and Th17 Cell Deficiency

(see Primer of Clinical Immunobiology «Chapter—T-helper cells»)

7 (DAF*0189:2009). Hyper-IgE syndrome results from a heterozygous loss-of-function mutation in Stat3, an intracellular signalling molecule distal to receptors for IL-6, IL-1 and IL-23. Predisposition to Candida and Staphylococcal infections results from deficiency of which of the following T-cell subsets?

A. Th1
B. Th2
C. Th17
D. Tfh

Answer C

Hyper-IgE syndrome is a rare genetic condition—I've personally only seen one case—resulting in susceptibility to Candida and Staphylococcal infections. But physiologically it's very interesting, so the key here is to understand the underlying principles of pathogenesis.

The question gives away all the clues: because STAT3 is a signalling molecule distal to receptors for IL-6, IL-1, and IL-23, a loss-of-function in this molecule will result in a diminished response to these cytokines; and all three are important in the biology of Th17 cells.

You will see from the following diagram that naive Th0 cells differentiate into Th17 cells under the influence of TGF-β together with IL-6 and IL-1, with further stabilisation of the phenotype by IL-23. So if you can't respond to these cytokines, Th17 cells fail to devel-op properly, leading to increased susceptibility to infections, particularly fungal infections.

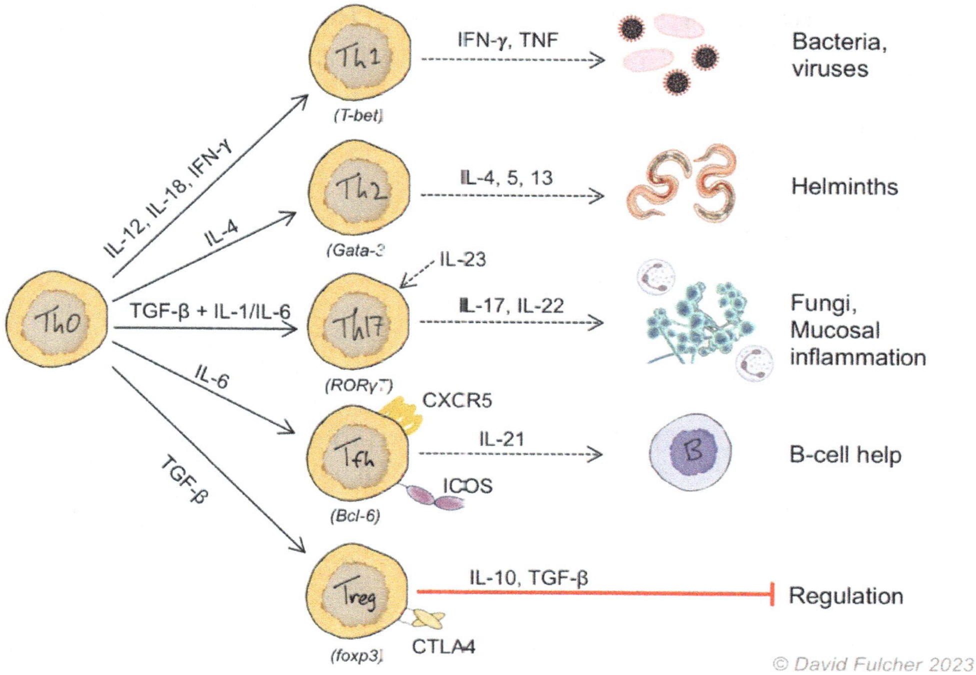

(Note in passing that there is also another genetic condition caused by a gain-of-function mutations in STAT3; this is also rare, and leads to early-onset lymphoproliferation and autoimmunity, particularly immune cytopenias, type 1 diabetes mellitus, polyarthritis, inflammatory bowel disease, and interstitial pneumonia.)

Question 8. Targeting T Cell Activation: The Role of Abatacept

(see Primer of Clinical Immunobiology «Chapter—How T-cells are Activated»)

8 (DAF*0192:2009). Which of the following potential treatments for lupus is aimed specifically at inhibiting T-cell activation?

A. Rituximab
B. Anti-BAFF (/BlyS)
C. Abatacept (CTLA4-Ig)
D. Anti-PD-1

Answer C

T cell activation requires two signals: signal 1, where the T cell receptor recognises peptide presented on HLA molecules; and signal 2, provided by B7 (on the APC) binding to CD28 (on the T cell). Abatacept is a fusion protein created by coupling soluble CTLA-4 with the Fc end of IgG. The CTLA-4 bit binds B7 molecules with high affinity, thereby blocking interaction with CD28 and preventing T cell activation.

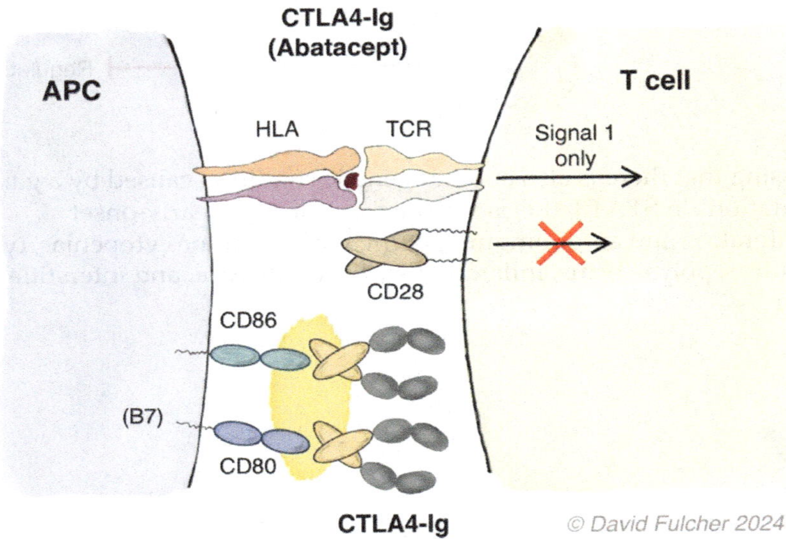

Let's look at the other choices:

- Rituximab targets CD20 and wipes out B cells. While B cells can indirectly influence T cell activation, rituximab is not specifically directed at T cells.
- Anti-BAFF (B cell activating factor): BAFF mainly supports B cell survival during the transitional stages of development, so again, anti-BAFF will have no direct effect on T cells.
- PD-1 is one of the targets of checkpoint inhibitors and its blockade acts to reinvigorate exhausted or anergic T cells, an effect utilised to great success in tumour immunology. In this context, the inhibitory signal that is otherwise conveyed by PD-L1/2 is blocked, allowing the T cell to activate and kill tumour cells.

But wait, there is a flipside to this: not all anti-PD-1 monoclonals act to block PD-1! Indeed, some actually activate PD-1, thus imparting a dominant inhibitory signal, 'turning off' the targeted T cell. Such a strategy has been successfully employed in autoimmune diseases such as rheumatoid artritis. So whilst 'D' has some truth to it, it is not as universally true as 'C'!.

> **I have been told by an ex-MCQ setter for the Royal Australian College of Physicians that a MCQ works best if several or all of the other options are correct / partially correct, and the candidate has to pick the one that is MOST correct!**

Question 9. Rituximab Effects in Autoimmune Disease

(see Primer of Clinical Immunobiology «Chapter—How B cells are Activated»)

9 (DAF*0194:2009). Following the successful use of Rituximab (anti-CD20) for pemphigus vulgaris, an autoimmune disease of the skin, all of the following changes would be expected EXCEPT:

A. Decrease in titres of autoantibodies
B. Decrease in tetanus-specific antibodies
C. Decrease in CD19+ cells (% lymphocytes)
D. Increase in CD3+ cells (% lymphocytes)

Answer B

Rituximab tends to affect autoantibody production more than antibodies against foreign antigens, an observation that suggests that autoantibodies are produced by short-lived plasmablasts which continue to express CD20 (and thus are susceptible to rituximab), whereas long-lived plasma cells, such as those that maintain pathogen-specific (and vaccine-specific) immunity lack CD20 expression and are unaffected by rituximab.

So decreased autoantibody levels would indeed be an expected outcome following rituximab given for autoimmune disease because of its effect on these autoreactive B cells, whereas a decrease in tetanus-specific antibodies is generally not seen.

With regard to the other two options:

- CD19+ lymphocyte proportions will decrease after rituximab, given that CD19 is a B-cell marker and B cells are removed by rituximab.
- CD3+ T cells will increase as a percentage of lymphocytes; this is not due to an actual rise in T cell numbers, but rather to the loss of B cells, leaving T cells (and for that matter NK cells) to fill the gap!

Now there is one caveat to the contention that 'anti-foreign' antibodies are generally preserved after rituximab treatment, and that is that some patients develop panhypogammaglobulinaemia; indeed rituximab is one of the major causes of secondary antibody deficiency. Obviously in those patients, levels of anti-foreign antibodies do decrease in parallel with the loss of total IgG.

Question 10. IVIG Use in Specific Antibody Deficiency

(see Primer of Clinical Immunobiology «Chapter—Antibody Deficiency»)

10 (DAF*0196:2009). Intravenous immunoglobulin is indicated in Australia for patients with a normal total IgG level, but who nevertheless have humoral immunodeficiency as evidenced by:

A. IgA deficiency
B. deficiency of both IgA and one or more IgG subclasses
C. Poor pneumococcal vaccine responses
D. low switched memory B cells

Answer C

We have to assume in this patient that there is clinical evidence of a 'humoral immunodeficiency' as a starting point, presumably with a convincing history of recurrent sinopulmonary infections. In that context, if the patient has normal immunoglobulins but fails to mount an adequate antibody response to pneumococcal vaccination or infection, then a diagnosis of specific antibody deficiency (SpAD) is appropriate, and may be an indication for accessing immunoglobulin replacement therapy (IgRT, IV or SC) in Australia. To get approval though, the patient first needs to have failed a trial of prophylactic antibiotics.

With regard to the other options:

- IgA deficiency is not an indication for IgRT.
- IgG subclass deficiency is also not an indication for IgRT; patients would need to demonstrate inadequate antibody responses to vaccination and thus be reclassified as SpAD.
- Although a low proportion of switched memory B cells is one of the diagnostic criteria for CVID, this diagnosis is excluded by virtue of having normal total IgG levels, and in any case, a reduction in switched memory B cells in itself is not an indication for IgRT.

Question 11. NK Cell Activation and Missing Self

(see Primer of Clinical Immunobiology «Chapter—Innate Cells»)

11 (DAF*0197:2009). NK cell killing of a target cell is most likely if the target cell:

A. expresses peptide antigen presented on HLA class II
B. expresses peptide antigen presented on HLA class I
C. down-modulates HLA class I
D. down-modulates HLA class II

Answer C

NK cells kill their target under two main circumstances: if the target is coated with IgG (termed 'Antibody-Dependent Cell-mediated Cytotoxicity', ADCC), or if the target has downregulated its HLA class I molecules.

To facilitate the killing of cells lacking HLA class I, NK cells possess two types of receptors, activating and inhibitory. A danger/ damage signal is expressed by the cell following infection or neoplastic transformation, and provides an initial stimulus to the NK cell's activating receptor, however if HLA class I molecules are expressed by that cell, the NK's inhibitory receptor will also be engaged, and no activation will occur. On the other hand, if HLA class I expression is lost ('missing self'), there will be no inhibitory signal, and the NK cell will deliver its lethal hit:

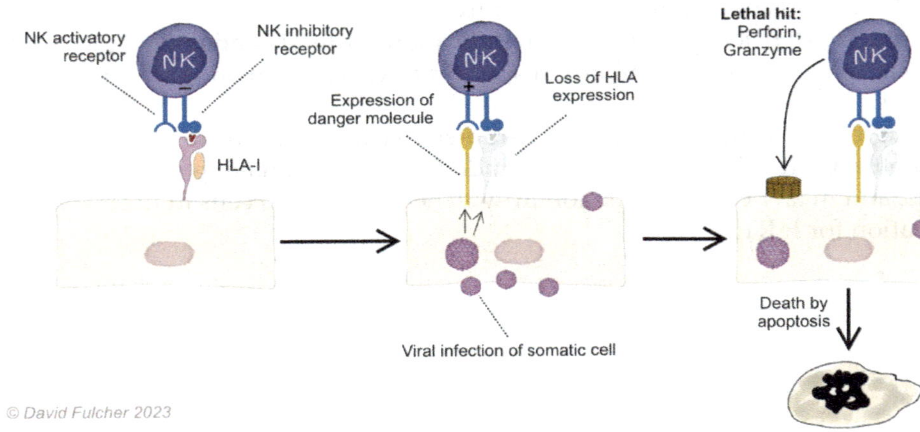

Question 12. TAP Deficiency and CD8 T Cell Development

(see Primer of Clinical Immunobiology «Chapter—How T cells are Made»)

12 (DAF*0201:2009). Deficiencies of Transporter-associated proteins TAP-1 and TAP-2 result in a deficiency of CD8 T cells, because they are essential for:

A. expression of HLA class I on thymic epithelial cells in primary lymphoid tissue
B. expression of HLA class II on thymic epithelial cells in primary lymphoid tissue
C. expression of HLA class I by antigen-presenting cells in secondary lymphoid tissue
D. CD8 T cell exodus from the thymus

Answer A

TAP deficiency is rare and one wouldn't expect your average general physician to know much about it, but it does illustrate an important principle of T-cell development. Immature thymocytes arrive in the thymus expressing neither CD4 nor CD8 ('double-negative'), then come to express both CD4 and CD8 ('double-positive'), finally emerging from the thymus as mature T cells, expressing either one or the other accessory molecule. The mechanism of this transition is explained in the Primer, but suffice it to say that if the thymocyte's TCR binds preferentially to HLA class I molecules on thymic epithelial cells, the cell downregulates CD4 and differentiates into a single-positive CD8 T cell, whereas if it binds to HLA class II molecules, CD8 is downregulated, and the cell emerges into a single-positive CD4 T cell.

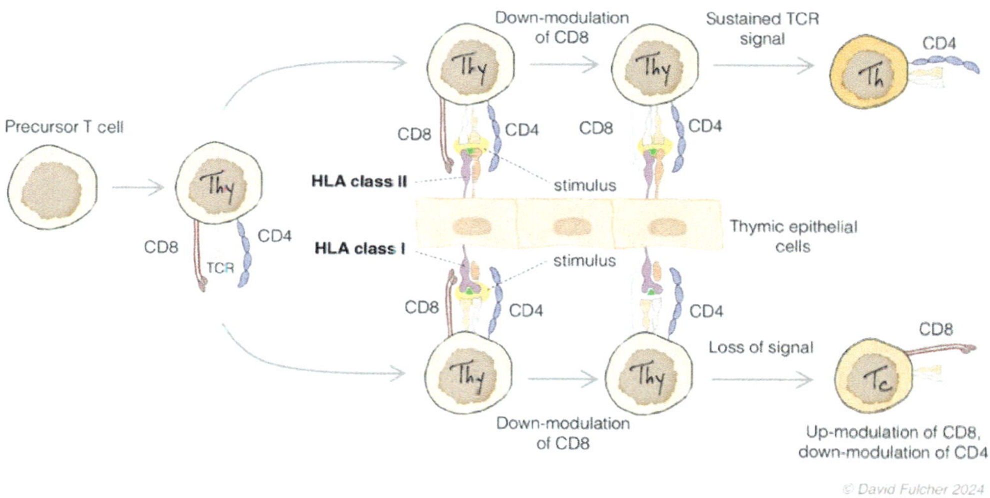

Now, HLA class I molecules comprise three components: the heavy chain, β-2 microglobulin, and the presented peptide, and all three need to be present for the complex to be surface-expressed. Given that TAP1 and TAP2 are essential for transporting peptides into the ER (where HLA class I molecules assemble), their absence means that peptide transport cannot occur and HLA class I molecules won't assemble. And absent HLA class I means that CD8-positive T cells have no pathway to selection.

In brief, a deficiency of TAP1 or TAP2 prevents normal HLA class I expression, leading to failure of CD8 single-positive differentiation.

With regard to other options:

- HLA class II expression is relevant only for CD4 T cell development, not CD8
- CD8 lineage commitment occurs in the primary, not secondary lymphoid tissue
- TAP proteins are for intracellular peptide transport, not thymocyte trafficking.

Question 13. T Cell-Independent Antibody Production

(see Primer of Clinical Immunobiology «Chapter—B-cell Activation»)

13 (DAF*0203:2009). Production of which of the following immunoglobulin isotypes is NOT T-cell dependent?

A. IgE
B. IgM
C. IgA
D. IgG

Answer B

Mature B cells that are activated without T cell help (for example, by certain microbial motifs or by mitogens), usually differentiate directly to plasmablasts which secrete IgM. B cells activated in such a 'T-independent' fashion are generally unable to undergo isotype switching necessary to produce other immunoglobulin classes.

By contrast, isotype switching, the process by which B cells switch from producing IgM to IgG, IgA, or IgE, almost always requires T cell help.

This is illustrated nicely by X-linked hyper-IgM syndrome, an 'inborn error of immunity' (IEI) resulting from deficiency of CD40. CD40L is a critical co-stimulatory molecule that mediates T-cell helper signals to B cells; when absent, B cells are unable to undergo isotype switching, affinity maturation nor memory B cell differentiation. Affected individuals can only make IgM, with marked deficiency of other immunoglobulin classes.

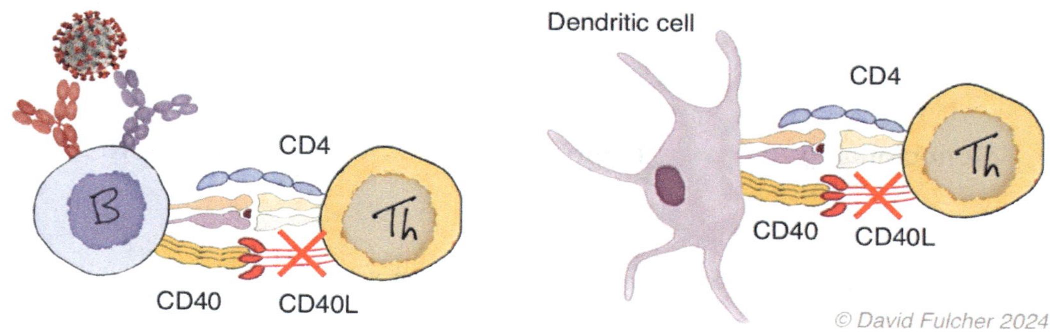

Question 14. Cryoglobulinemia Types and Rheumatoid Factor

(see Primer of Clinical Immunobiology «Chapter—Immunoglobulin Disorders»)

14 (DAF*0206:2009). A patient with a peripheral vasculitic rash is found to have cryoglobulinaemia. The cryoglobulin is found to consist of a monoclonal IgM paraprotein and polyclonal IgG. Which is the following tests is most likely to be positive?

A. RF
B. SSA
C. dsDNA
D. HBV-DNA

Answer A

The cryoglobulin constituents described in the question correspond to type 2 cryoglobulinemia. Recall the classification:

- Type 1: Monoclonal immunoglobulin only (usually IgG or IgM); no rheumatoid factor.
- Type 2: Monoclonal IgM (with rheumatoid factor activity) binding to polyclonal IgG.
- Type 3: Polyclonal IgM (with rheumatoid factor activity) binding to polyclonal IgG.

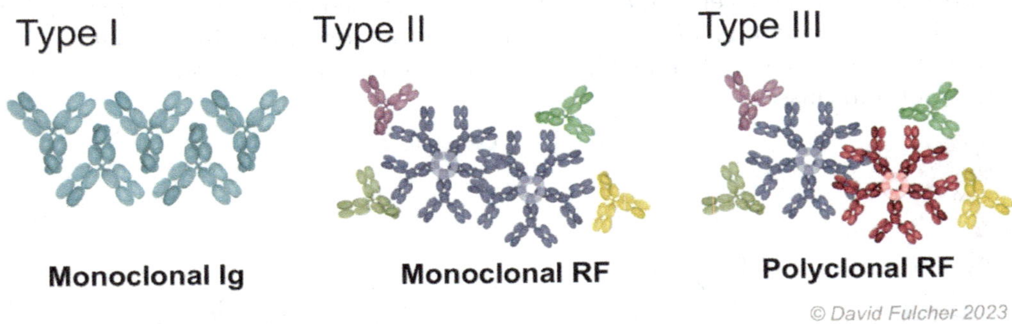

So all type 2 cryoglobulinemia patients will have a positive RF, making this the correct answer.

Antibodies to SSA and dsDNA may be found in a subset of rheumatoid factor-positive autoimmune patients with type 3 cryoglobulinemia, but here the cryoglobulin constituents are polyclonal. Finally, hepatitis C is commonly associated with type 2 (and some-times type 3) cryoglobulinemia, but not hepatitis B.

Question 15. Plasmacytoid Dendritic Cells and Viral Interferon Response

(see Primer of Clinical Immunobiology «Chapter—Innate Cells»)

15 (DAF*0209:2009). Which of the following cell types is most responsible for the interferon response to viral infections?

A. Monocytes
B. Macrophages
C. Plasmacytoid dendritic cells
D. B cells

Answer C

The major claim to fame for plasmacytoid dendritic cells (pDCs) is their role in sensing virus and for churning out buckets of type I interferon! To do so, pDCs express innate PRRs specialised for detection of various viral PAMPs, especially nucleic acids. The type I interferons they release act to induce an antiviral state in exposed cells (locally and systemically) by creating a cellular environment that is resistant to viral replication.

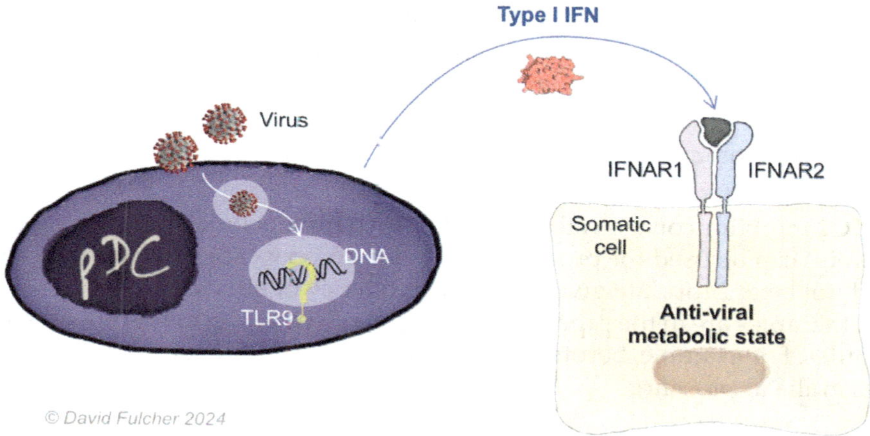

Many other cells (including the ones in this question) can also produce type I interferons to some degree, but pDCs are by far the most specialised and potent source following viral infection.

Question 16. Hereditary Angioedema Management

(see Primer of Clinical Immunobiology «Chapter—Complement Disorders»)

16 (DAF*0211:2009). A 26 year old woman presents to casualty for the fourth time in two years with spontaneous development of lip swelling. Labs have previously shown absent serum C4, and reduced levels of C1 inhibitor and function. Which is the most appropriate long-term management:

A. Intermittent steroids for acute attacks
B. Weekly infusions of C1 inhibitor concentrate
C. Fortnightly injections of lanadelumab
D. Icatibant to be taken at the earliest sign of severe swelling

Answer D

Recurrent angioedema of the lip with absent C4 and reduced C1 esterase inhibitor levels is consistent with a diagnosis of hereditary angioedema (HAE) type I. The question then is how best to manage a patient who has relatively infrequent, non-life-threatening episodes, and specifically whether to commence prophylactic treatment.

Given the low attack frequency and absence of life-threatening events, the best management option here is intermittent icatibant. Icatibant is a bradykinin receptor antagonist that patients can self-administer at the earliest onset of swelling, thus usually avoiding the ED.

With regard to the other options:

- Steroids are ineffective in HAE.
- While C1 inhibitor concentrate or lanadelumab (monoclonal antibody against kallikrein) can be used for prophylaxis, these are very expensive options and would not be appropriate given the low frequency of attacks.

Daily oral therapies are in the pipeline, offering excellent preventive strategies and good safety profiles. For example, berotralstat is available but very expensive (and not on the PBS in Australia at present).

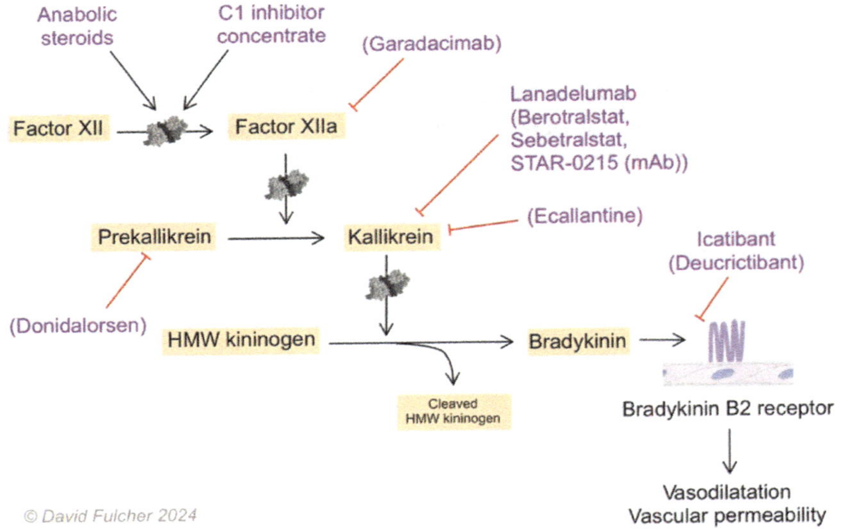

Question 17. Classic Mendelian Inheritance and Immunodeficiencies

(see Primer of Clinical Immunobiology «Chapter—Antibody Deficiency», «Chapter—Innate Cells», «Chapter—How B cells are Activated», «Chapter—B-cell Development»)

17 (DAF*0213:2009). Classic Mendelian inheritance is characteristic of all the following immunodeficiencies EXCEPT:

A. Common Variable Immunodeficiency
B. Hyper IgM syndrome
C. Chronic granulomatous disease
D. Bruton's agammaglobulinaemia

Answer A

CVID is genetically complex. There are a wide variety of gene mutations that have been associated with CVID, which themselves follow Mendelian inheritance (though some with incomplete penetrance), but importantly (at the time of writing) more that half of CVID cases completely lack any clear genetic explanation. Therefore, among the listed conditions in the question, CVID is the least consistent with classical Mendelian inheritance, making it the correct answer.

Here are some of the known CVID-associated mutations, and these would of course each follow their own Mendelian inheritance pattern. Note also the middle column of other genetic immunodeficiencies which can give rise to a phenotype identical to CVID when the mutation is less damaging (hypomorphic).

'Common'	Hypomorphic mutations found in other genetic immunodeficiencies	Rare
TACI*	RAG1/RAG2 (SCID)	BAFF-R
CTLA4	JAK3 (SCID)	ICOS
LRBA	ZAP70 (SCID)	KMT2D
PIK3CD	CARD11 (CID)	IKZF1
PIK3R1	CD40L (HIGM)	TCF3
NFKB1	AIDCA (HIGM)	BACH2
NFKB2	BTK (XLA)	LIG4
	CD19 (B- hypogammaglobulinaemia)	STXBP2
	BLNK (B- hypogammaglobulinaemia)	LIG1
	UNC93B1 (HSV encephalitis)	RTEL1
	XIAP (XLP)	FOXP3
	CD27 (XLP)	PMS2
	WAS (Wiskott-Aldrich)	TINF2
	STAT3 (HIES)	DNMT3B
	DOCK8 (HIES)	ZBTB24
	CXCR4 (WHIM)	IL12RB1
		IKBKB
		IL12B
		IL21R
		ITCH
*Susceptibility gene only: probably requires other mutations for expression		

On the other hand, classic Mendelian inheritance is a feature of all the other listed immunodeficiencies in the question:

- The most common form of hyper-IgM syndrome (CD40L deficiency) is X-linked, whilst the other even rarer ones are autosomal recessive.
- The most common form of chronic granulomatous disease is also X-linked (specifically NOX2 (or gp91) deficiency) whilst the others (p22, p67, p47 and rac) are autosomal recessive.
- Bruton's agammaglobulinemia (XLA) is X-linked.

Question 18. Complement Factor C2 Deficiency

(see Primer of Clinical Immunobiology «Chapter—Complement Disorders»)

18 (DAF*0215:2009). A deficiency of C2 is most likely to result in:

A. Recurrent attacks of angioedema
B. Neisseria infections
C. Mesangiocapillary glomerulonephritis type II (dense deposit disease)
D. Lupus-like disease

Answer D

C2 is part of the classical complement pathway, and its deficiency is associated with immune complex-mediated diseases, such as lupus. The pathophysiology involves impaired clearance of immune complexes (antigen-antibody aggregates), since complement facilitates this process; thus without efficient clearance, these complexes deposit in tissues and trigger inflammation. Patients with C2 deficiency are also prone to recurrent pyogenic infections, but among the listed options, lupus-like disease is the most appropriate.

Classical pathway	Manifestations
C1, C4, C2	Autoimmune disorders e.g. SLE, glomerulonephritis Recurrent pyogenic infections (e.g. *S. Pneumoniae* and *H. Influenza* type b).
Common pathway	
C3	Severe pyogenic infections (a/a) Autoimmune disorders e.g. SLE, glomerulonephritis
Alternate pathway	
Properdin Factor D	Neisseria infections
Terminal pathway	
C5, 6, 7, 8, 9	Neisseria infections

With regard to the other possibilities:

- Recurrent angioedema occurs in C1 inhibitor deficiency.
- *Neisseria* infections are associated with defects in the terminal complement pathway (membrane attack complex), as well as properdin deficiency.
- Dense deposit disease is usually driven by C3 nephritic factor which drives alternative pathway dysregulation.

Question 19. Paroxysmal Nocturnal Hemoglobinuria (PNH)

(see Primer of Clinical Immunobiology «Chapter—Complement Disorders»)

19 (DAF*0216:2009). Paroxysmal Nocturnal Haemaglobinuria occurs as a result of

A. A genetic lack of transmembrane-bound complement control proteins
B. An acquired lack of transmembrane-bound complement control proteins
C. A genetic lack of GPI-linked complement control proteins
D. An acquired lack of GPI-linked complement control proteins

Answer D

Paroxysmal nocturnal hemoglobinuria (PNH) is an acquired disorder. It results from a somatic mutation in haematopoietic stem cells, most commonly affecting the PIGA gene. This mutation disrupts the synthesis of GPI (glycosyl-phosphatidyl-inositol) anchors, which are needed to attach several key complement regulatory proteins to the surface of blood cells.

GPI-linked complement regulatory proteins (such as CD55 and CD59) help protect cells (e.g. red blood cells) from complement-mediated damage; without the GPI anchor, these protective proteins are absent from the cell surface, and red cells become susceptible to uncontrolled complement deposition leading to intravascular haemolysis.

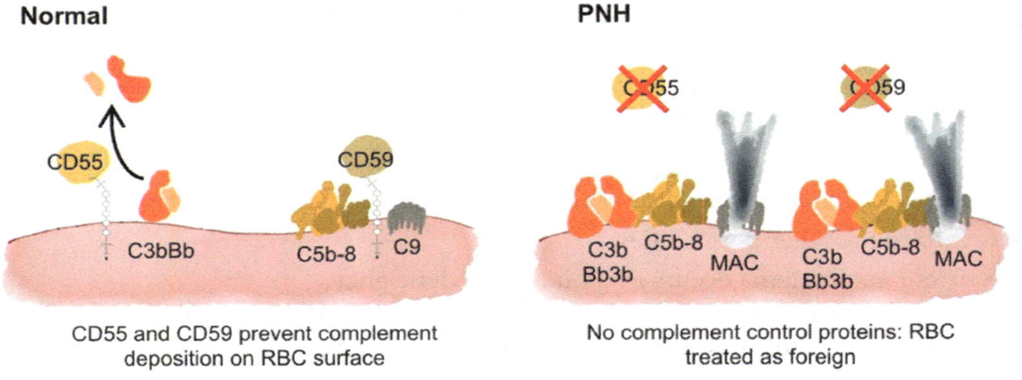

© David Fulcher 2023

Regarding the other options, PNH does not arise due to a defect in a transmembrane protein, nor is it inherited.

Question 20. ACE Inhibitor-Induced Angioedema

(see Primer of Clinical Immunobiology «Chapter—Non-Atopic Allergic Disease»)

20. (DAF*0217:2025). A 72 year old man presents to emergency with acute swelling of the tongue. He has a history of hypertension, ischaemic heart disease, and type 2 diabetes. Which of the following is the most likely cause for the tongue swelling?

A. Amyloidosis
B. C1 inhibitor deficiency (hereditary angioedema)
C. Adverse drug reaction to perindopril
D. Idiopathic (spontaneous) angioedema

Answer C

The salient points to note in this question are that this is the first presentation of acute onset tongue angioedema in a 72 year old man with hypertension. We see combination a great deal in ED, and the most likely cause is ACE inhibitor-induced angioedema.

> **I would be rung in the middle of the night from ED at least once a month with this same type of presentation, and the first question I always asked was, "does he have a history of hypertension", and the second, "is he on an ACE inhibitor". Yes. Yes....**

This is a well-recognised and potentially life-threatening side effect of this class of drug. Patients will often give a history of more minor swellings, and preceding ACEI-related cough seems to be a risk factor. The angioedema may also affect the airway, and in-deed the tongue swelling can be severe enough to cause asphyxia.

With regard to the other options:

- Amyloidosis can involve the tongue, but amyloid progresses slowly, over months to years, in contrast to the acute presentation here.
- Hereditary angioedema is theoretically possible but highly improbable as a first presentation at this age.
- Idiopathic angioedema is plausible; if the patient were NOT on an ACE inhibitor it is certainly the next diagnostic consideration.

Question 21. Monoclonal Antibodies Isotype and Function

(see Primer of Clinical Immunobiology «Chapter—Immunoglobulins»)

21 (DAF*0226:2010). A monoclonal antibody with the LEAST chance of causing immune activation will be of which isotype?

A. IgG1
B. IgG2
C. IgG3
D. IgG4

Answer D

Fc-mediated immune functions, such as complement activation or binding to Fc receptors on cells, is weakest in the case of IgG4 by comparison to the other isotypes. Consider the list of such functions in the Table below, and note the many "—" entries for IgG4! As a result, IgG4 is considered a 'blocking antibody'.

IgG subclasses: Properties	IgG1	IgG2	IgG3	IgG4
Heavy chain	γ1	γ2	γ3	γ4
Approximate proportion in serum	66%	23%	7%	4%
Serum level (g/L)	9	3	1	0.5
Half-life	21	21	7.5	21
Approximate molecular weight (kDa)	146	146	165	146
Hinge (number of amino acids)	15	12	62	12
ADCC	+++	+/− −	+++	+/− −
Phagocytosis	+	+	+	+/− −
Classical C' pathway activation	+	+/−	+++	−
Complement-mediated cytotoxicity	++	+/−	++	−
Neonatal FcR binding	+	+	+/−	+
Cell interactions	FcγR on APCs			

Isotype-related function is an important consideration in design of therapeutic monoclonal antibodies; if the goal is to kill a target cell, strong Fc effector functions are required. On the other hand, if the aim is simply to block a receptor and not engage down-stream immune pathways, IgG4 is the one to go for!

Question 22. Chronic Mucocutaneous Candidiasis and Th17 Defects

(see Primer of Clinical Immunobiology «Chapter—T-helper cells»)

22. This photo shows candida-related onycholysis of the toenails. Inherited deficiency of which protein could explain this condition in a 16 year old female?

A. Mannose-binding lectin (MBL)
B. Gamma-common chain
C. CD40L (CD154)
D. IL-17 receptor

Answer D

The image shows candida-related onycholysis and skin infection in a 16-year-old female with a chronic, inherited susceptibility to fungal infections. To answer this question, recall that protection against candida is largely the role of Th17 cells, so anything impairing the Th17 pathway is a possible cause for CMC. Genetic defects that may affect the Th17 pathway can involve the cardinal cytokines IL-17 and IL-22 themselves, the signalling components driving their production, or defects in their receptors/receptor signalling pathways:

Pathogenetic group	Defect
Genetic	IL-17RA IL-17F ACT1 Dectin-1 CARD9 Stat1-gain of function
Autoimmune (APS-1/APECED, thymoma)	Anti-IL-17 Anti-IL-22

The most common cause of inherited CMC, and the diagnosis that was found in this patient, is a gain-of-function mutation in STAT1. This mutation promotes Th1 differentiation at the expense of Th17, since STAT1 is distal to cytokines which drive Th1 differentiation.

With regard to the other options:

- Mannose-binding lectin deficiency is mild and not clinically significant in almost all cases; it certainly could not be responsible for the severe phenotype in this patient.
- Common gamma chain deficiency leads to SCID, which typically presents much earlier, and with a wide range of other opportunistic pathogens.
- CD40 ligand deficiency (X-linked hyper-IgM syndrome) does make the patient susceptibility to 'T-cell type' infections. but the course is typically characterised by antibody-deficiency infections, neutropenia, and PJP, not isolated chronic cutaneous fungal infections here.

Question 23. Pathogenic vs Marker Autoantibodies

(see Primer of Clinical Immunobiology «Chapter—Autoimmunity»)

23 (DAF*0236:2010). An autoantibody is shown to be pathogenic if all the following apply EXCEPT:

A. It causes disease in neonates of affected mothers
B. It is positive in almost all patients with an autoimmune disease
C. Levels fluctuate with disease activity
D. It causes disease on injection into mice

Answer B

How do we determine whether an autoantibody is truly pathogenic rather than just a marker of disease? This is clinically important: if an antibody is causing damage, removing it may improve outcomes (e.g., through rituximab, IVIG, or plasma exchange).

Well, this question emphasises some of the important considerations. A pathogenic antibody should cause disease in neonates born to affected mothers. It should cause disease when transferred to a third party, typically an experimental animal. Its levels should fluctuate with disease activity; less direct, but still supportive.

BUT, merely being present in affected patients does not prove pathogenicity. Such an autoantibody may simply be a disease marker, a bystander in the autoimmune process, whilst not necessarily causing disease.

Question 24. IgA Deficiency

(see Primer of Clinical Immunobiology «Chapter—Antibody Deficiency»)

24 (DAF*0238:2010). Which of the following statements about IgA deficiency is INcorrect?

A. It is typically asymptomatic
B. It may be genetically associated with common variable immunodeficiency
C. It may present as a transfusion reaction
D. It may result in false positive coeliac serology

Answer D

Perhaps this is a bit of a trick question, because if you read it quickly, you might not have noticed that it says 'false positive' rather than 'false negative' coeliac serology. But it's important to understand the logic.

IgA-based serological assays such as anti-tTG and endomysial antibody (EMA) rely on the patient having IgA! So if IgA is absent, the test result will be falsely negative; in such cases, IgG-based tests such as anti-gliadin IgG are needed, notwithstanding their lower specificity.

With regard to the other options, IgA deficiency (IgAD)...

- ...is typically asymptomatic. Many individuals with IgA deficiency are identified incidentally (indeed often on coeliac screening!) and remain well.
- ...can be genetically associated with CVID and be present in family members of patients with this condition; some IgAD patients may even evolve into a CVID phenotype over time.
- ...can present as a transfusion reaction. This happens because some IgAD patients develop anti-IgA antibodies, such that transfusing them with blood products containing traces of IgA can result in an immune complex-mediated reaction, occasionally severe.

Question 25. V(D)J Recombination and RAG Proteins

(see Primer of Clinical Immunobiology «Chapter—B-cell development», «Chapter—T-cell development»)

25 (DAF*0243:2010). Binding of Recombinase Activating Gene (RAG) proteins to V, D and J sequences is critical for the development of:

A. just T cells
B. just B cells
C. T & B cells
D. all lymphocytes

Answer C

RAG proteins (RAG1 and RAG2) are essential for the development of lymphocytes that require V-D-J recombination, namely B cells and T cells. RAG proteins mediate the recombination of variable (V), diversity (D), and joining (J) gene segments, essential steps in generating a diverse B and T cell receptor repertoire.

RAG1 and RAG2 work together to initiate the cutting and joining of DNA segments that form the antigen receptor genes. Without RAG proteins, neither B cells nor T cells can develop functional receptors, leading to severe immunodeficiency (as seen in RAG-deficient SCID).

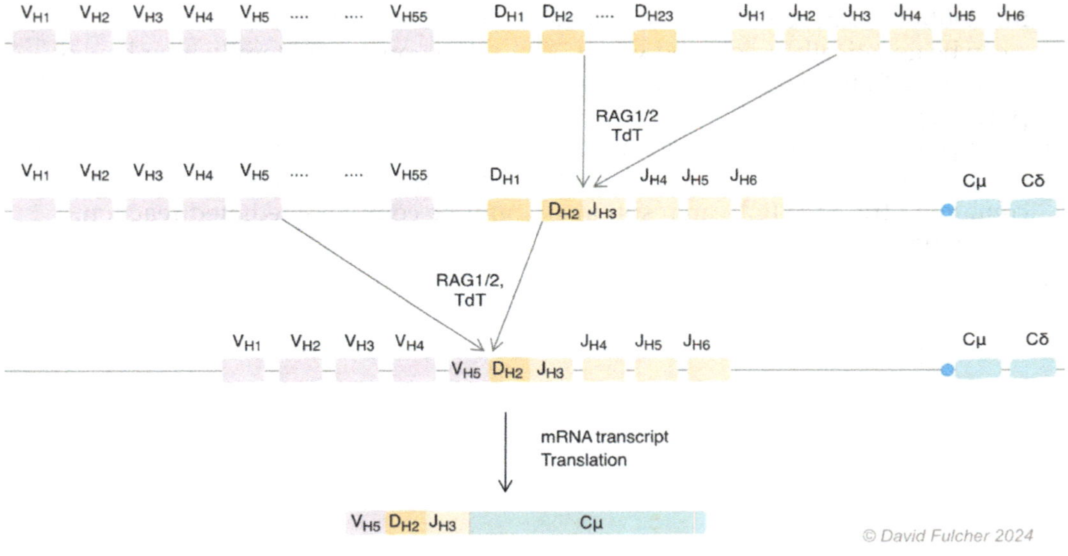

In contrast, natural killer (NK) cells and innate lymphoid cells (ILCs) do not have specific antigen receptors and thus are not dependent on V-D-J recombination for their development.

Question 26. SARS-CoV-2 and the Inflammasome

(see Primer of Clinical Immunobiology «Chapter—Innate Receptors»)

26 (DAF*0919:2025). Viroporin 3a is a protein of the SARS-CoV-2 virus which can activate NLRP3. The effect of this activation would be release of:

A. IL-2
B. IL-4
C. IL-1
D. TGF-beta

Answer C

Activating the NLRP3 inflammasome, and inflammasomes in general, leads to the release of interleukin-1β (IL-1β), a potent pro-inflammatory cytokine. Recall that NLRP3 is a member of the NOD-like receptor (NLR) family; some NLR monomers can assemble into large multiprotein complexes called inflammasomes, which when activated send a strong activation signal to the cell. In this example, the NLRP3 inflammasome is activated after binding viroporin 3a, which allows it to cleave procaspase-1 into active caspase-1, which in turn processes pro-interleukin-1β into active IL-1β.

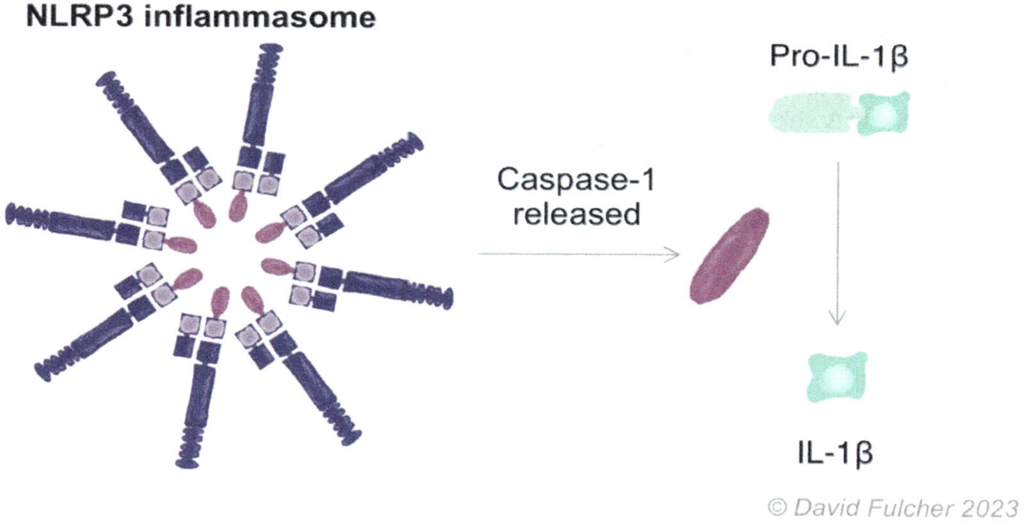

The inflammasome pathway may thus contribute to acute inflammatory responses, such as the cytokine storm observed in severe SARS-CoV-2 infection.

In this series of MCQs, we have now come across two examples of NLRP3 activators, one (here) relating to innate protection against SARS-CoV-2 virus, and earlier in gout, where urate crystals activate the same inflammasome. Either way, IL-1β-mediated inflammation is the result!

Question 27. Distinguishing CVID from Secondary Causes

(see Primer of Clinical Immunobiology «Chapter—Antibody Deficiency»)

27 (DAF*0920:2025). Which of the following would negate a diagnosis of Common Variable Immunodeficiency (CVID)?

A. Normal IgA
B. Normal memory B cells
C. Thymoma
D. Normal dynamic vaccine responses

Answer C

This question emphasises a key concept in clinical immunology: before one can diagnose CVID, secondary causes of antibody deficiency must be excluded. One such secondary cause is thymoma; although quite rare, the association of thymoma with hypogammaglobulinaemia is termed Good's syndrome. So if a patient has thymoma (or other malignancies such as lymphoma or chronic lymphocytic leukemia), and is found to have low immunoglobulins, then antibody deficiency is likely to be secondary. In such cases, a diagnosis of CVID should not be made.

In contrast:

- Normal IgA levels do not rule out CVID: ESID (European Society of Immunodeficiency) criteria classify hypogamma patients who have normal IgA/IgM as "possible CVID" rather than "probable CVID". In other words, an absence of IgA strengthens the diagnosis, but a normal level does not exclude it.
- Normal memory B cells are seen in about a third of CVID patients. While having low switched memory B cells supports a CVID diagnosis, having normal memory B cell proportions does not exclude it.
- Normal vaccine responses are atypical but nevertheless possible in CVID.

So the key point is, if another cause of antibody deficiency is found (e.g. thymoma), CVID should not be diagnosed.

Question 28. Targets of B-cell depleting monoclonals

(see Primer of Clinical Immunobiology «Chapter—How B cells are Activated»)

28 (DAF*0921:2025). Inebilizumab is a humanized IgG1κ monoclonal antibody against CD19 that is effective in the treatment of myasthenia gravis. Which of the following B-lineage cells would be targeted by inebilizumab but unaffected by rituximab?

A. Mature B cells
B. Immature B cells
C. Memory B cells
D. Pro-B cells

Answer D

Current B-cell depleting drugs generally target either CD19 (e.g. inebilizumab) or CD20 (e.g. rituximab), both pan-B-cell surface markers, so the question is asking about the differences in terms of the maturational stages of B-lineage cells which will be targeted by these two respective agents?

The most important difference is that CD19 is expressed early in B-cell development, and CD20, slightly later:

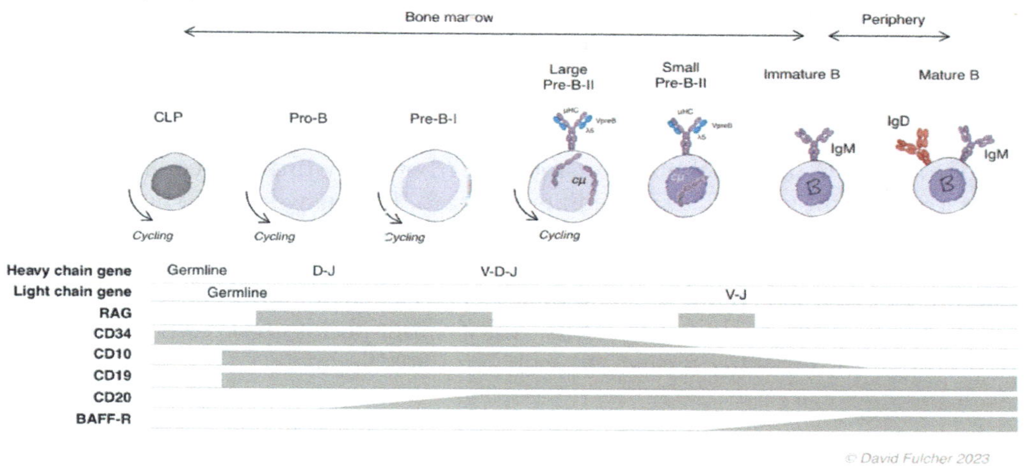

This means that CD19-targeting antibodies like inebilizumab will eliminate those earlier stages of B-lineage development, whereas these stages are unaffected by rituximab. Later B-cell maturation stages, such as immature B, mature B, memory B, and most plasmablasts, express both CD19 and CD20. Finally, plasma cells generally express neither surface marker.

There is of course a downside to wiping out these early CD19-expressing B-lineage cells, namely the potential to disrupt normal B-cell regeneration. Thus, whilst rituximab therapy leaves these early stages alone, this is not the case for anti-CD19 monoclonals; treated patients therefore risk losing the capacity to generate new B cells, thus hindering the ability to make new antibody responses in the future.

Question 29. Indications for B-cell depleting monoclonals

(see Primer of Clinical Immunobiology «Chapter—How B cells are Activated», «Chapter—How B cells are Made»)

29 (DAF*0922:2025). B-cell depleting monoclonal antibodies have been shown to be effective in all the following EXCEPT:

A. Myasthenia Gravis
B. IgG4 disease
C. Ankylosing spondylitis
D. Multiple sclerosis

Answer C

B-cell-depleting monoclonal antibodies have demonstrated efficacy in several autoimmune diseases, specifically those where B cells are intrinsically involved in immunopathogenesis. Such would include:

- Myasthenia gravis, a classical autoantibody-mediated disease. There's ample evidence that B cell-depleting therapies like rituximab and inebilizumab, are effective.
- IgG4-related disease. The pathogenesis of this condition remains a little enigmatic, but since IgG4 is made by B cells, it makes sense that B-cell depleting therapies would be effective. Indeed multiple studies have shown this (although that doesn't mean that these therapeutics are necessarily licensed for this indication!).
- Multiple sclerosis. Anti-CD20 monoclonals such as ocrelizumab and ofatumumab are widely used and effective in MS, but the demonstration of benefit posed a bit of a conundrum to immunopathophysiologists (there's a word!), since MS was hitherto thought to be a T-cell-mediated disease. The answer might lie in MS being a multifactorial process involving autoantigen presentation by B cells, B-cell release of pro-inflammatory cytokines, and formation of ectopic lymphoid follicles in the meninges, all of which might be impacted by B-cell depletion.

So the exception is ankylosing spondylitis. Ankylosing spondylitis is considered to be a Th17-driven disorder, with little evidence for B cell involvement.

Here is the table from the Primer listing some of the important autoimmune conditions where B cell depletion is of clinical benefit:

Effective	Maybe effective	Probably ineffective
Acquired angioedema with C1-inhibitor deficiency ANCA-associated vasculitis Autoimmune haemolytic anaemia Behçet's disease Bullous pemphigoid Castleman's disease Cryoglobulinemia Goodpasture's disease IgG4-related disease Immune thrombocytopenia Juvenile idiopathic arthritis Multiple sclerosis (esp. relapsing-remitting) Myasthenia gravis Nephrotic syndrome Neuromyelitis optica spectrum disorders (NMOSD) Pemphigus Rheumatoid arthritis Spondyloarthropathy Systemic sclerosis.	Membranous nephropathy Primary Sjögren's syndrome Graves' disease	Antiphospholipid syndrome Autoimmune hepatitis IgA nephropathy Inflammatory myositis Primary-progressive multiple sclerosis Systemic lupus erythematosus (SLE) Ulcerative colitis.

Question 30. Tezepelumab and Allergic Disease

(see Primer of Clinical Immunobiology «Chapter—Pathophysiology of Allergic Disease»)

30 (DAF*0923:2025). Tezepelumab is an anti-TSLP monoclonal antibody which is effective in the treatment of severe chronic rhinosinusitis with nasal polyposis and asthma. This therapeutic acts at the level of:

A. Innate Lymphoid Cells (ILCs)
B. Th1 cells
C. Th2 cells
D. Cytotoxic T cells

Answer A

Chronic rhinosinusitis with nasal polyposis (CRSwNP) is a notoriously challenging condition to manage, often requiring repeated surgical interventions, so biologic therapies like tezepelumab offer promising alternatives.

At the immunological level, TSLP itself is an epithelium-derived 'alarmin', part of the early innate response to environmental stress-es. Along with IL-25 and IL-33, it acts on group 2 innate lymphoid cells (ILC2s) whose job it is to produce type 2 cytokines such as IL-5 and IL-13, which promote eosinophilic inflammation and drive the development of Th2 responses, the hallmark of allergic disease.

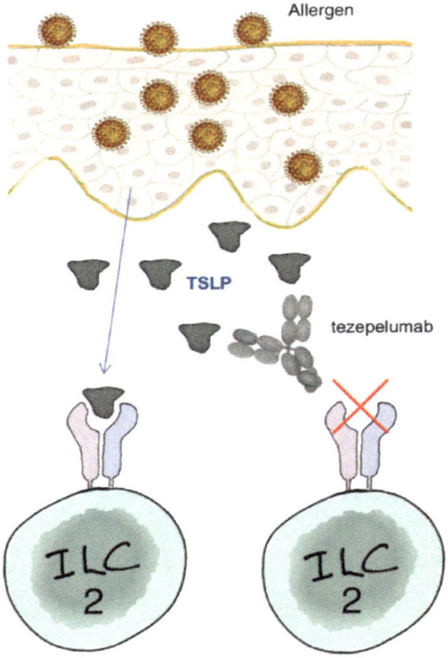

© David Fulcher 2024

By blocking TSLP, tezepelumab interrupts this upstream signalling cascade, reducing activation of ILC2s and, in turn, type 2 inflammation. Given its upstream position in the allergic pathway, tezepelumab may also prove beneficial in the future for the treatment of a broader range of other type 2-related diseases than asthma. The precise role of TSLP in pseudo-allergic conditions such as CRSwNP isn't clear, but its inhibition seems an effective therapeutic strategy.

Question 31. Telitacicept and B-Cell Maturation

(see Primer of Clinical Immunobiology «Chapter—How B cells are Made»)

31 (DAF*0924:2025). Telitacicept is a fusion protein of TACI linked to Fc end of IgG which has been trialled in lupus. Which aspect of B-cell biology does this inhibitor target:

A. B-cell generation in the marrow
B. The passage of B-cells through the transitional stage
C. B cell activation by T cells
D. Establishment of antibody forming cells in bone marrow niches

Answer B

Now's a time to remind yourself of the rather complex system of BAFF/receptor combinations, but here's a quick revision:

- BAFF binds to three receptors on B-lineage cells: BAFF receptor (BAFF-R), TACI, and BCMA.
- APRIL can also both bind to TACI and BCMA (but not BAFF-R)

Telitacicept is in effect a soluble form of TACI made to have a decent half-life by coupling it to the Fc end of IgG. It binds BAFF and APRIL thereby preventing their action on the above three receptors, which are (mostly) on B cells.

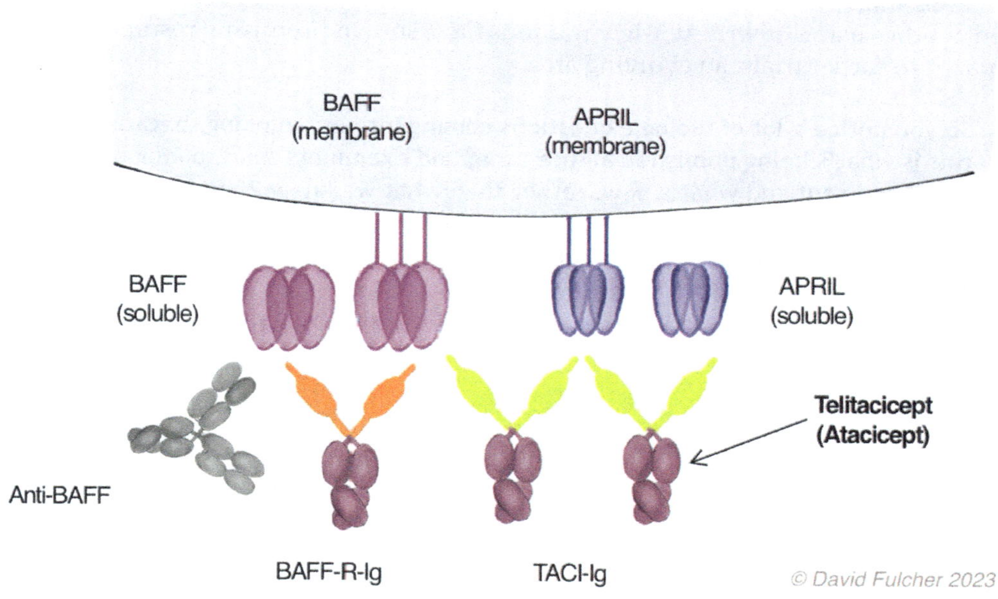

Now recall that immature B cells pass through transitional stages to become fully mature B cells, a process that is dependent on BAFF signalling. Telitacicept will therefore block these critical survival signals and inhibit B cell maturation and survival:

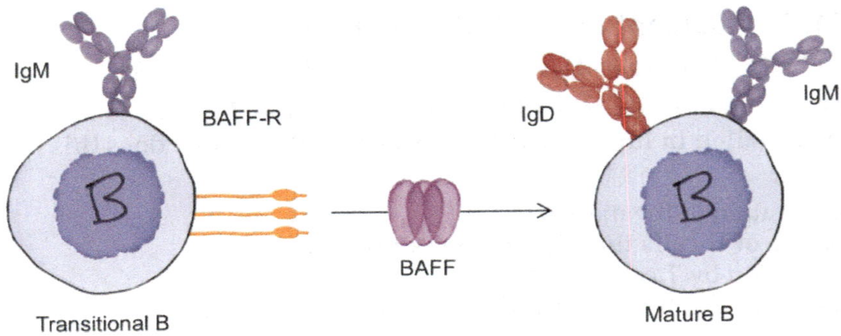

With regard to the other options:

- B cell generation in the marrow is not reliant on BAFF.
- B cell activation by T cells is not a function of BAFF, but rather performed by interactions mostly involving CD40 to CD40L, and ICOS to ICOSL.
- Telitacicept may affect plasmablasts or short-lived antibody-secreting cells as a result of its blockage of the interaction of BAFF and APRIL on BCMA, but it does not meaningfully target the establishment or persistence of long-lived plasma cells in bone marrow niches.

Recent studies using other TACI-Ig variations have shown promising results in lupus compared to earlier trials, an evolving area.

> **So you notice a lot of biologic questions coming into my thinking, because this is what's being published all the time, and examiners tend to look at what's current and what's new, rather than what we knew 20 years ago. This is why I try to teach in the course the basic fundamental underlying immunobiology, which allows us to understand these new therapeutics and why they work.**

Question 32. JAK Inhibitors and Giant Cell Arteritis

(see Primer of Clinical Immunobiology «Chapter—More T Cells and T-Cell Therapeutics»)

32 (DAF*0925:2025). Upadacitinib is of potential benefit in the management of patients with giant cell arteritis, mostly likely because of its effect on:

A. Blocking IL-6
B. Blocking the IL-6 receptor
C. Blocking the downstream signalling of the IL-6 receptor
D. Blocking release of IL-6 from macrophages

Answer C

Upadacitinib is a JAK inhibitor that has shown potential benefit in the treatment of inflammatory and allergic disorders. Recent publications have extended its uses to include giant cell arteritis (GCA), probably due to its impact on interleukin-6 (IL-6) signalling, given that IL-6 is a key cytokine in the pathogenesis of GCA.

When IL-6 binds to its receptor, it activates JAK1, which then phosphorylates STAT molecules that drive pro-inflammatory gene expression. Upadacitinib inhibits JAK1, thereby disrupting this downstream signal cascade, and reducing IL-6-mediated inflammation.

JAK1 is also used by other cytokine receptors, meaning that upadacitinib (and indeed other JAK inhibitors) can block multiple cytokine pathways simultaneously and therefore have potential roles in the management of other autoimmune and allergic diseases.

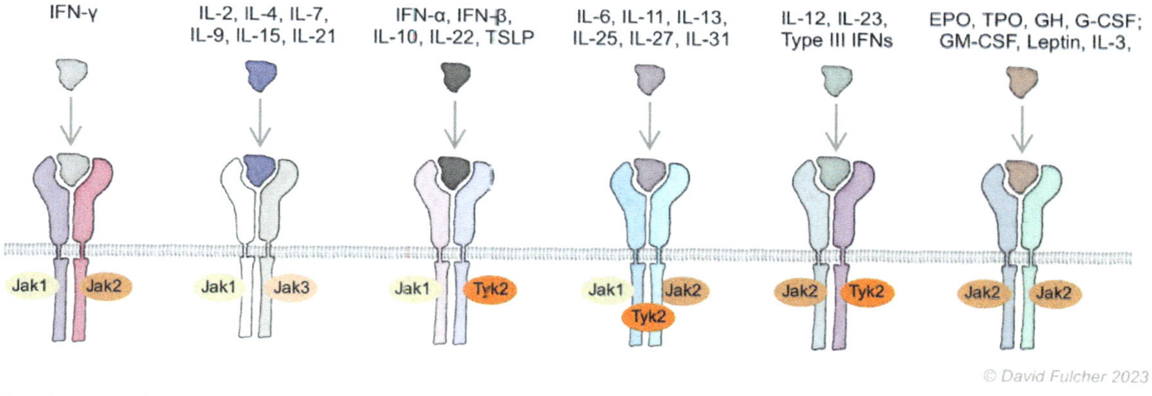

With regard to the other options, upadacitinib:

- does not block IL-6 directly (that's what siltuximab does).
- does not block the IL-6 receptor (as tocilizumab does).
- does not inhibit IL-6 release (can't think of anything that would do that!).

Question 33. Allergies and Persistence with Age

(see Primer of Clinical Immunobiology «Chapter—Atopic Allergic Disease», «Chapter—Non-Atopic Allergic Disease»)

33 (DAF*0926:2025). Hypersensitivity reactions to which of the following allergens is LEAST likely to decline with age?

A. Penicillin
B. Peanut
C. Milk
D. Soy

Answer B

Let's eliminate the easier options first. Milk and soy allergies are primarily restricted to paediatrics; most affected infants outgrow these allergies during toddlerhood or early childhood.

The other two options are not so obvious. Whilst "penicillin allergy" often reflects a mislabelling or non-IgE-mediated reaction, even in true IgE-mediated cases, sensitisation tends to fade over time; studies show that over 80% of patients lose their penicillin sensitivity after 10 years without re-exposure.

Contrast that with peanut allergy; only a minority 'outgrow' peanut allergy, estimated to be maybe ~20–25% of children. So in most cases, peanut allergy persists into adulthood.

Question 34. T Cells and Oral Tolerance

(see Primer of Clinical Immunobiology «Chapter — T-cell Tolerance»)

34 (DAF*0927:2025). T-cells play a pivotal role in maintenance of tolerance to food antigens. The T-cell subset most associated with this role produces which of the following cytokines?

A. Interferon-alpha
B. Interleukin-22
C. Interleukin-10
D. Interleukin-gamma

Answer C

Regulatory T cells are central to the maintenance of tolerance to food and other environmental antigens, particularly a subset known as peripheral regulatory T cells (pTregs). These cells exert their suppressive effects through the production of immunoregulatory cytokines, the key ones being IL-10 and TGF-β. These cytokines play critical roles in downregulating immune responses, maintaining oral tolerance, and preventing inappropriate reactions to food antigens.

A note about the other options:

- Interferon-alpha (IFN-α) is involved in innate antiviral responses, not tolerance.
- Interleukin-22 (IL-22) is produced by Th17 cells, which are involved in barrier immunity and inflammation.
- Interferon-gamma (IFN-γ) is produced by Th1 cells, promoting macrophage activation and cellular immunity.

Question 35. Complement Inhibitors and IgA Nephropathy

(see Primer of Clinical Immunobiology «Chapter—Complement Disorders»)

35 (DAF*0928:2025). Iptacopan is an oral inhibitor of Complement Factor B that has beneficial effects in IgA nephropathy. This is consistent with the observation that IgA nephropathy is a disorder driven by:

A. the lectin pathway of complement activation
B. the classical pathway of complement activation
C. the alternate pathway of complement activation
D. the membrane attack complex

Answer C

The question is basically asking, 'which complement pathway involves factor B?' And the answer is the alternate pathway.

So a quick revision of the alternate complement pathway. It starts with spontaneous generation of active C3b, which then binds factor B, forming the C3 convertase C3bBb. C3bBb can generate more C3b as well as initiate assembly of the membrane attack complex. But by inhibiting factor B, iptacopan blocks C3bBb formation and thus subsequent downstream effects:

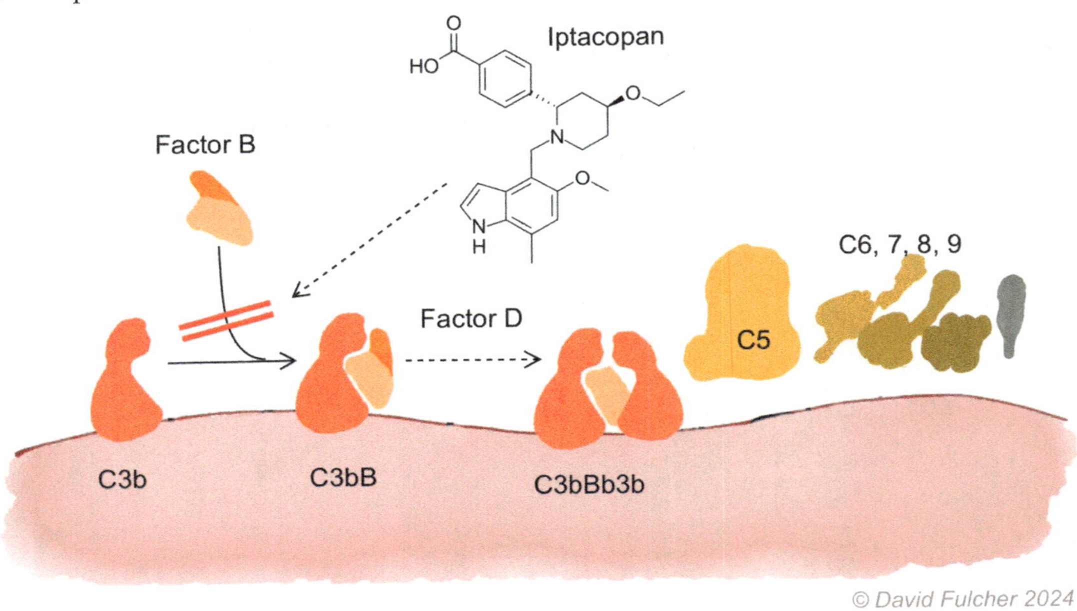

There is good evidence that the alternate pathway is active in IgAN, and recent publications (2025) suggest that iptacopan is indeed of therapeutic benefit.

> In a previous life, I used to look at the immunofluorescence of diagnostic renal biopsies, noting the staining pattern seen in IgA nephropathy, where immune deposits typically contain bulk IgA and C3, but not C1q. The absence of C1q reinforced the conclusion that the complement classical pathway was not involved here. Indeed if mesangial C1q were present,

you'd be thinking more along the lines of lupus nephritis (e.g., mesangial proliferative lupus), rather than IgA nephropathy.

Question 36. Bispecific T Cell Engagers: Blinatumomab

(see Primer of Clinical Immunobiology «Chapter—More T cells and T cell Therapeutics»)

36 (DAF*0929:2025). Blinatumomab is a bispecific, single-chain molecule that targets CD19 and CD3 ("BITE"). Such a therapeutic could be of potential therapeutic utility in which of the following conditions?

A. T-cell lymphoblastic lymphoma
B. NK-LGL leukaemia
C. Acute myeloid leukaemia
D. B-cell acute lymphoblastic leukaemia

Answer D

Blinatumomab is not a complete antibody, but a complex of two variable region antibody fragments designed to bring T cells into proximity with B cells, tricking the T cells into killing B cell targets. To do this:

- one arm of the BiTE molecule binds CD3 on T cells; remember that CD3 is the part of the T-cell receptor complex, and transmits the activation signal.
- the other arm of the BiTE binds CD19 which is expressed by B-lineage cells (normal and malignant).
- The drug therefore clusters the T cell with the B cell, whilst also sending an activation signal to the T cell.
- The T cell then does what cytotoxic T cells do, and delivers a lethal hit on the B cell.

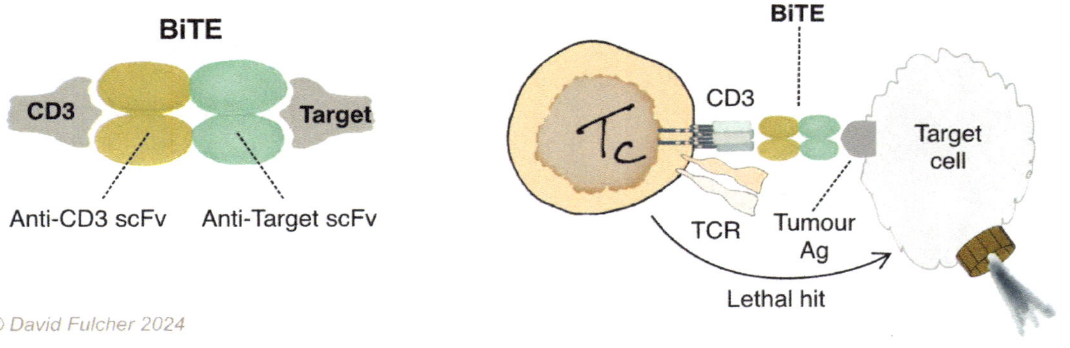

© David Fulcher 2024

Blinatumomab should therefore kill any CD19-expressing cells, and in most cases has been used to treat B-cell malignancies. So coming back to the question, the only B-lineage condition listed is B-ALL.

It's worth stressing that blinatumomab depletes all CD19+ cells, including normal B cells, a reasonable trade-off in exchange for therapeutic efficacy for aggressive B cell malignancies. But unsurprisingly it causes progressive hypogammaglobulinaemia meaning that monitoring of IgG levels during and after treatment is mandatory, with consideration given to IgRT.

Question 37. NK Cell Deficiency and Herpes Infections

(see Primer of Clinical Immunobiology «Chapter—Innate Cells»)

37 (DAF*0930:2025). NK cell deficiency should be suspected in any patient with early onset:

A. Herpes virus infections
B. Pneumocystis pneumonia
C. sinopulmonary infections
D. Neisseria infections

Answer A

Natural killer (NK) cell deficiency is extremely rare and usually presents to paediatricians. It should be suspected in any patient presenting with early-onset or severe herpesvirus infections, especially if other causes have been excluded.

But adult physicians should recall that NK cells have an important role in control of herpesviruses. This is likely due to the knack of herpesviruses in downregulating HLA class I expression on infected cells, thus avoiding recognition by cytotoxic T cells. Nevertheless, such down-modulation licenses the activation of NK cells by the 'missing self' pathway of NK cell action (see Question 11).

With regard to the other options:

- Pneumocystis infections suggest general T cell deficiency (think AIDS or SCID). That's not to say that T-cell deficient patients do not have a problem with herpesvirus infection, just that other 'T-cell-type' infections tend to dominate the clinical presentation.
- Recurrent sinopulmonary infections point to B cell or antibody deficiency.
- *Neisseria* infections are associated with terminal complement pathway defects.

Question 38. Therapeutics Targeting Cytokines

(see Primer of Clinical Immunobiology «Chapter—T-helper cells»)

38 (DAF*0931:2025). Which of the following therapeutics targets only a single cytokine or cytokine action?

A. Upadacitinib (Jakinib)
B. Dupilumab (anti-IL4R-alpha)
C. Ustekinumab (anti-IL-12p40)
D. Benralizumab (anti-IL5R-alpha)

Answer D

This is a challenging question, as it requires an understanding of various cytokines, their receptors, and the actions of the specific therapeutics.

Let's work through it:

- Ubitacitinib is a JAK1 inhibitor. As discussed previously (Question 32), JAK1 is used by multiple cytokine receptors, so this drug inhibits multiple cytokine pathways. Definitely not the answer!
- Dupilumab targets the IL-4 receptor α, which is used by both the IL-4 and IL-13 receptors. So, while it looks like an IL-4 path-way blocker, it functionally blocks the action of both these cytokines:

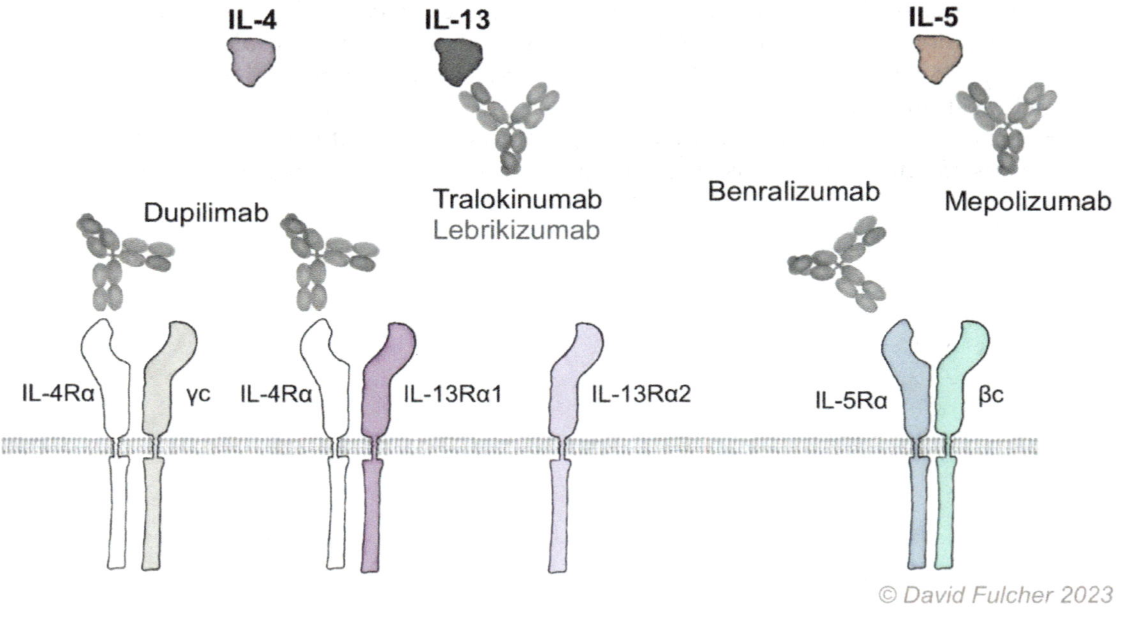

- Ustekinumab is an antibody against IL-12 p40, a subunit shared by IL-12 and IL-23, and so blocks both cytokines:

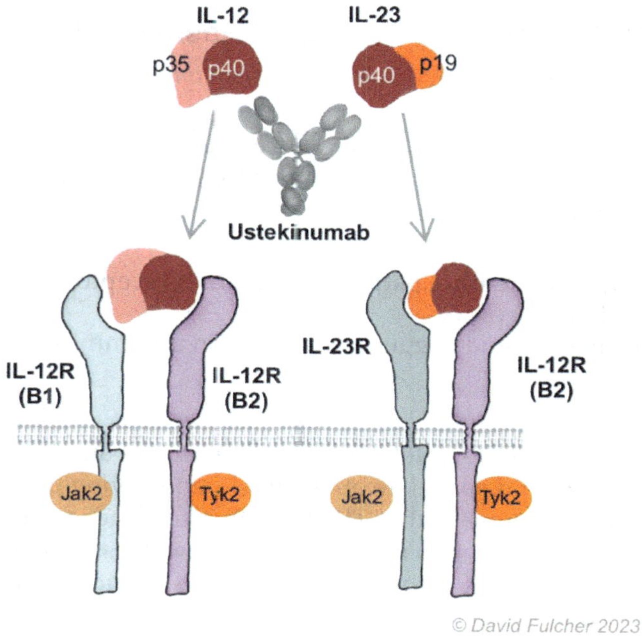

- Benralizumab on the other hand targets the IL-5 receptor α subunit, which is only used by IL-5. It does not affect any other cytokines, making this the correct answer.

Question 39. TREX1 Deficiency and Cytoplasmic DNA

(see Primer of Clinical Immunobiology «Chapter—Autoimmunity», «Chapter—Innate Immunity»)

39 (DAF*0932:2025). TREX1 deficiency is a rare cause of an early-onset lupus-like disorder. Its deficiency allows accumulation of DNA in the cytoplasm, which likely contributes to lupus pathogenesis by:

A. activating anti-DNA-specific B-cells on cell lysis
B. binding to innate nucleotide pattern-recognition receptors
C. binding to toll-like receptors
D. stimulating formation and triggering of NLRs in inflammasomes

Answer B

Although only a handful of TREX1 deficiency cases exist globally, the question here asks you to work through and speculate on the likely underlying pathogenesis. In the scenario provided, the key detail is that TREX1 deficiency leads to accumulation of DNA in the cytoplasm. The question then becomes: "how does cytoplasmic DNA contribute to the development of autoimmunity in lupus?" And the answer is that cytoplasmic DNA is recognized by intracellular sensors, such as cGAS, which signal to induce type I interferon production, a central feature of lupus pathogenesis.

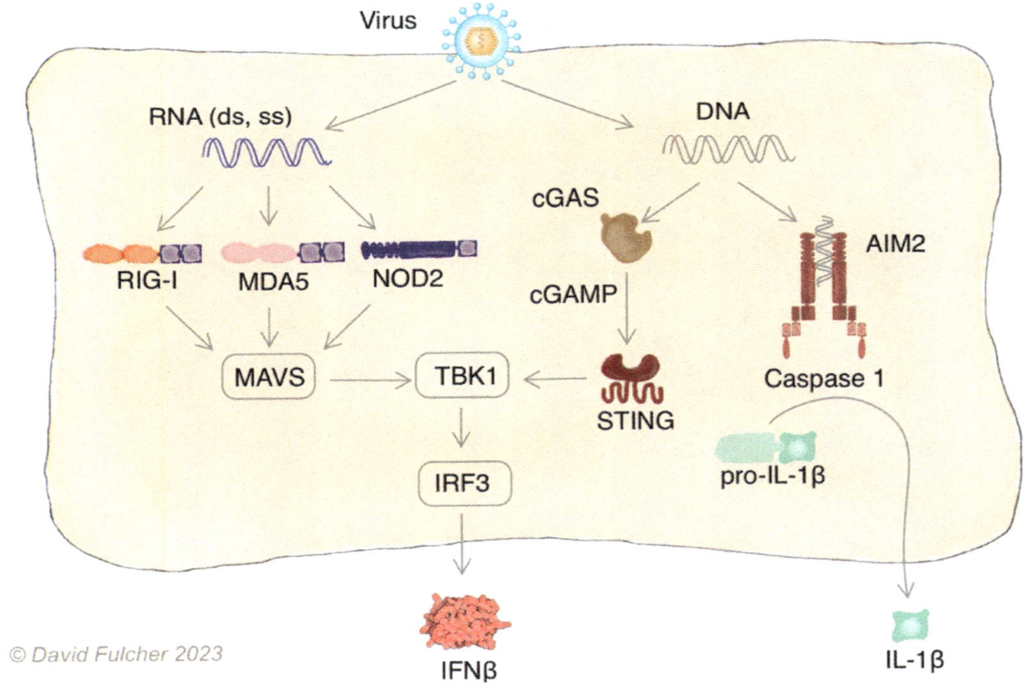

Let's consider the other options:

- Activation of DNA-specific B cells on cell lysis is unlikely to explain pathogenesis; although extracellular DNA can activate B cells, either via the BCR of anti-DNA-specific B cells or by engaging TLRs after phagocytosis, the same phenomenon must be happening all the time after necrosis of any random cell under whatever circumstance, i.e. there would be nothing specific about TREX1 deficiency.

- Binding to toll-like receptors is unlikely because TLRs are located either on the cell surface or within endosomes, not in the cytoplasm, so they won't sense accumulation of cytoplasmic DNA.
- Some NOD-like receptors can detect cytosolic nucleotides (NOD2 for example), but here it is acting as pattern recognition receptor, not as an inflammasome. The only DNA sensor that forms an inflammasome is AIM2, which is not an NLR!

So, in summary, TREX1 deficiency leads to accumulation of cytoplasmic DNA, which engages nucleotide-sensing receptors like cGAS, resulting in interferon production and contributing to lupus-like autoimmunity.

Question 40. T Cell Activation

(see Primer of Clinical Immunobiology «Chapter—How T cells are Activated»)

40 (DAF*0933:2025). In terms of T-cell activation, which of the following surface molecules is the 'odd one out'?

A. LAG-3
B. CTLA4
C. CD28
D. PD-1

Answer C

This is quick one, and hopefully you were all over it! The concept is important though and worth reviewing. In a nutshell:

- LAG-3 = co-inhibitory molecule
- CTLA-4 = co-inhibitory molecule
- PD-1 = co-inhibitory molecule
- CD28 = co-stimulatory molecule

So the odd one out is CD28, because it is the only co-stimulatory receptor in the list. The others all serve to inhibit T cell activation, thereby maintaining peripheral tolerance.

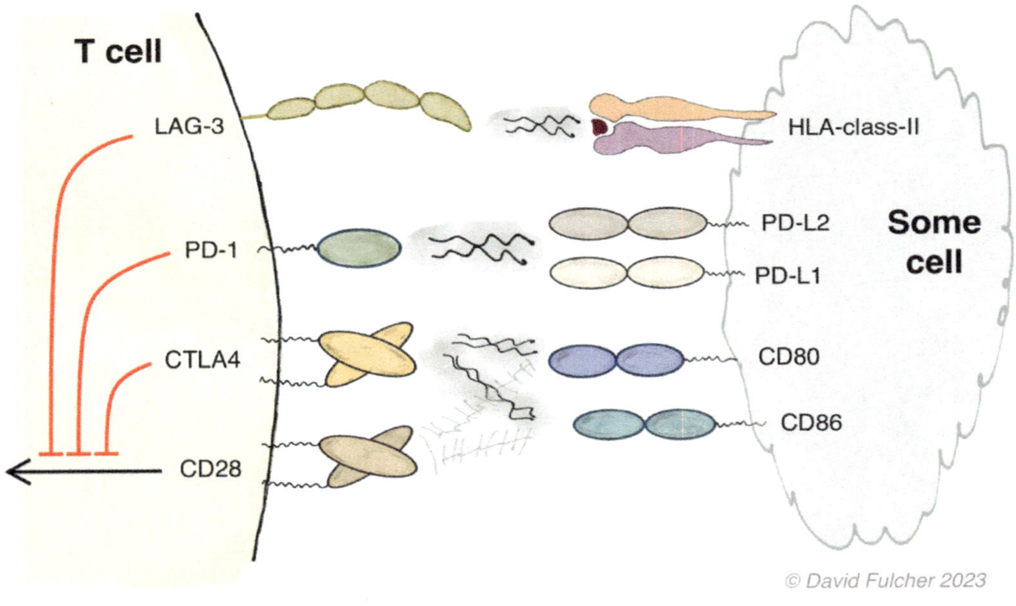

Question 41. T-Follicular Helper Cells and B Cell Differentiation

(see Primer of Clinical Immunobiology «Chapter—T-helper cells»)

41 (DAF*0934:2025). Which of the following functions is NOT characteristic of T-follicular helper cells?

A. B-cell differentiation to short-lived plasma cells
B. Production of high affinity antibodies by B cells
C. Class-switch recombination in B cells
D. Differentiation of naive to memory B cells

Answer A

Tfh cells are specialized CD4-helper T cells with a very focused purpose: to help B cells produce high-quality, high-affinity antibodies. They reside mostly in the germinal centres of secondary lymphoid organs, where they provide the signals required for:

- Isotype switching (to IgG, A, E)
- Affinity maturation
- Memory B cell differentiation
- Long-lived plasma cell differentiation

In contrast, short-lived plasma cells are generated outside the germinal centre, typically early in the immune response, where they produce low-affinity IgM. They are not the product of the germinal centre (where Tfh cells mostly reside), although T-cell help is usually required for their activation. Short-lived plasma cells serve a useful immediate function by provide that early IgM surge of protection.

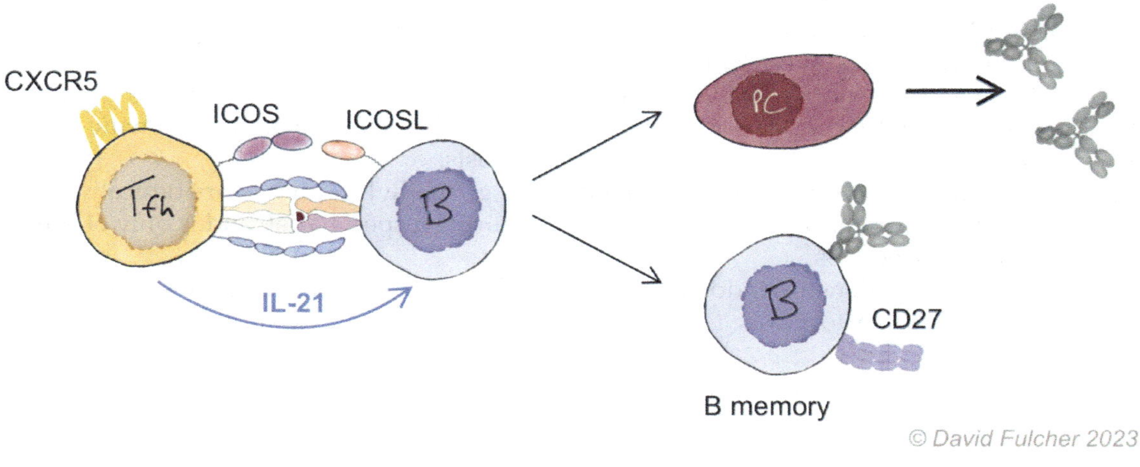

So in summary, Tfh cells are all about high-quality, long-term humoral immunity, and their key characteristics are:

- expression of CXCR5, which guides them into the B-cell follicle
- expression of ICOS, for B-cell co-stimulation
- production of IL-21, which drives B-cell proliferation, differentiation, and class switching

Question 42. Mastocytosis and Diagnostic Testing

(see Primer of Clinical Immunobiology «Chapter—Non-Atopic Allergic Disease»)

42 (DAF*0935:2025). An 18-year-old male with a long-standing history of presumed food allergies and recurrent anaphylaxis was referred for evaluation. Since infancy, he had experienced wheezing, rhinitis, and skin lesions, with episodes of flush-ing, angioedema, and presyncope often attributed to food triggers. Despite extensive dietary restrictions and treatments, including omalizumab, his symptoms persisted. Physical examination revealed widespread keratosis pilaris and nodular skin lesions exhibiting Darier's sign. Serum basal tryptase was 23 ng/mL (NR<11).

Which of the following would be least likely to confirm the clinical diagnosis?

A. Skin biopsy
B. Bone marrow biopsy for KIT D816V mutation
C. 24-hour urinary histamine
D. Serum chromogranin A level

Answer D

Although this patient was investigated and treated for food allergy, the presence of rash was a clue that this was not a very satisfactory explanation. After some delay, Darier's sign (whealing and erythema after the skin is gently scratched) was documented, a characteristic finding in mastocytosis. Further investigation confirmed an elevated serum tryptase, consistent with, but not proof of, the clinical diagnosis.

To establish mastocytosis, confirmatory tests are needed, and these may include:

- Skin biopsy, which should show aggregates of mast cells. Skin biopsy is a relatively straightforward investigation and often highly informative.
- Bone marrow biopsy to assess mast cell burden and to facilitate genetic testing, particularly for the KIT D816V mutation, which is the most common somatic mutation associated with systemic mastocytosis.
- 24-hour urinary histamine or metabolites (e.g., methylhistamine) are typically elevated, but this test is more complex, requiring a special diet, and is only offered in specialised labs.

On the other hand, neuroendocrine tumour markers are not relevant in this context and not elevated in mastocytosis.

Question 43. T Cell Precursors in the Thymus

(see Primer of Clinical Immunobiology «Chapter—How T cells are Made»)

43 (DAF*0936:2025). The earliest T-cell progenitor cells entering the thymus express:

A. both CD4 and CD8
B. CD4 only
C. CD8 only
D. neither CD4 nor CD8

Answer D

T cell progenitors move from the bone marrow to the thymus, entering as immature thymocytes, ready to begin T cell maturation and selection. At this stage they express neither CD4 nor CD8, and are therefore known as double-negative thymocytes (DN = $CD4^-$/ $CD8^-$), and with further differentiation, they progress to double-positive thymocytes (DP = $CD4^+$/$CD8^+$).

They then must commit to one of the two T-cell lineages dependent on the type of HLA with which their T-cell Receptor (TCR) can interact; they become $CD4^+$ helper T cells if they recognise HLA class II, or $CD8^+$ cytotoxic T cells if they recognise HLA class I (see Question 12).

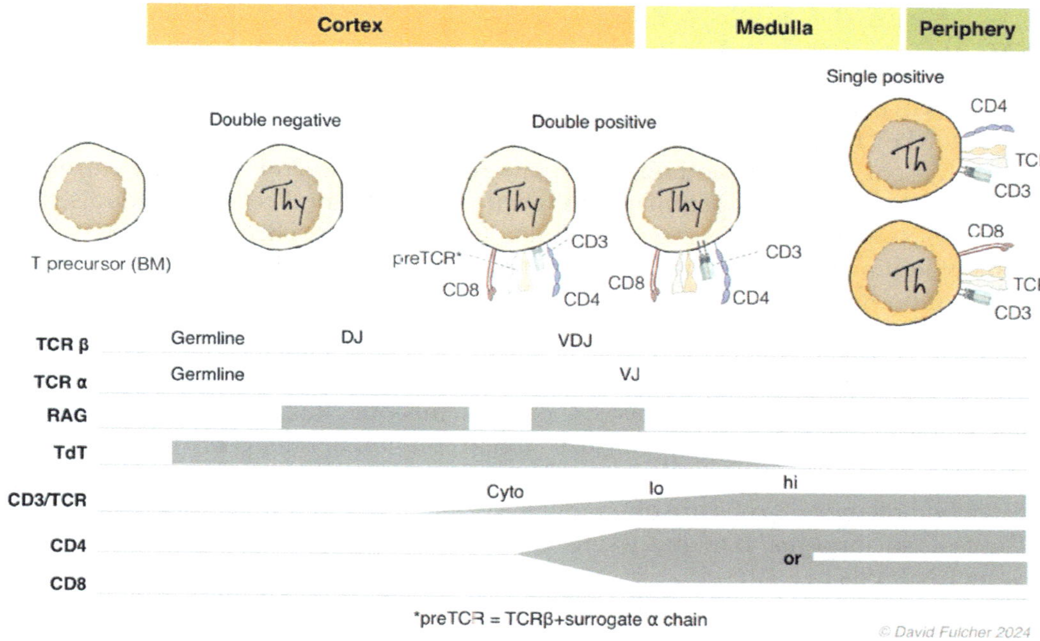

Question 44. CVID vs X-Linked Agammaglobulinemia

(see Primer of Clinical Immunobiology «Chapter—Immunoglobulin Deficiency»)

44 (DAF*0937:2025). A 40-year-old male with a history of recurrent respiratory infections since age 5 was diagnosed with common variable immunodeficiency (CVID) at age 23, when he presented with severe obstructive lung disease and markedly low immunoglobulin levels (IgG <0.2 (NR>7g/L), IgA <0.2 (NR>0.7g/L), and IgM 0.22 (NR>1.4 g/L)).

Despite immunoglobulin replacement therapy, which reduced infection frequency, he continued to suffer from pulmonary insufficiency. Physical exam showed pulmonary rhonchi, and chest imaging revealed hyperexpanded lungs. Pulmonary function was severely impaired (FEV1 at 24.3% predicted). Flow cytometry of lymphocytes revealed 77% CD3 cells, CD19 0%, and NK cells 20.3%. Family history was negative for immunodeficiency.

Which of the following tests would be most relevant to confirm or refute CVID as the diagnosis?

A. Tetanus toxoid antibody response after vaccination
B. Memory B cell studies
C. CT chest
D. Btk gene sequencing

Answer D

A diagnosis of common variable immunodeficiency (CVID) might sound reasonable at first, but the key finding here comes from flow cytometry, specifically that CD19+ B cells were absent. Notwithstanding the negative family history, this clinical picture (male, early onset, panhypogammaglobulinaemia, and absent B cells) should raise strong suspicion of X-linked agammaglobulinaemia (XLA), aka Bruton's agammaglobulinaemia. Although CVID and XLA can overlap clinically, distinguishing them has important implications, especially for genetic counselling.

So in the end, the question boils down to which test helps confirm or refute CVID (over XLA)? Let's work through the options:

- Tetanus toxoid antibody responses: CVID patients often fail to respond to vaccines, but some can still mount a response, so this result will neither confirm nor refute the diagnosis.
- Switched memory B cells are often low in CVID (~70%), but this is not invariably the case, so abnormal B memory cells would not help; of course in this case, there were no B cells to analyse!
- CT chest will show structural damage but adds nothing to distinguishing CVID from XLA.
- BTK gene sequencing is the crucial test. If a pathogenic mutation in BTK is found, the diagnosis is XLA, not CVID.

Unfortunately the discovery of hypogammaglobulinaemia as an explanation for recurrent infection in this patient was substantially delayed, resulting in nearly 20 years of recurrent pulmonary infections and irreversible lung damage. This is exactly the scenario that internists need to identify so that treatment with immunoglobulin replacement therapy (IgRT) can be promptly commenced and lung damage prevented.

Question 45. Immune Thrombocytopenia (ITP) in CVID

(see Primer of Clinical Immunobiology «Chapter—Immunoglobulin Deficiency»)

45 (DAF*0938:2025). A 65 year old woman was referred following the discovery of pan-hypogammaglobulinaemia on investigation for recurrent sinus and chest infections (IgG 0.6 g/l; IgA 0.16 g/l; IgM 0.44 g/l). Examination showed low-grade generalised lymphadenopathy. Other causes of immunodeficiency were excluded and a diagnosis of CVID made. IVIG was commenced and the sinus and chest symptoms improved.

Eleven months after the diagnosis she presented with a flat purpuric rash affecting her legs and blisters in her mouth. The next most relevant test is:

A. skin autoantibodies
B. measurement of switched memory B cells
C. platelet count
D. mouth biopsy

Answer C

A diagnosis of CVID seems reasonable in this middle-aged woman with panhypogammaglobulinaemia and lymphadenopathy, although we are not told about other confirmatory tests. When she subsequently presents with a flat purpuric rash, a clinical diagnosis of immune thrombocytopenia should immediately spring to mind, because patients with CVID have an increased risk of autoimmune disease, particularly immune cytopenias such as ITP. So the next most appropriate next test will be platelet count.

With regard to the other options:

- autoimmune blistering diseases such as pemphigus or pemphigoid, for which skin autoantibodies would be the next test, are rare (in CVID and in general!), and do not present with purpura.
- Switched memory B cells are usually low in CVID, but this finding would not be useful for establishing an explanation for this new complication.
- Mouth biopsy might be indicated if there were mucosal lesions or blistering but not the case here.

This case illustrates the 'variable' nature of CVID. While it may present solely with infections, many patients exhibit additional features such as:

- autoimmunity, particularly cytopenias (e.g., ITP, autoimmune haemolytic anaemia)
- lymphoproliferation (e.g. lymphadenopathy, splenomegaly)
- enteropathy.

In fact, there's a known subset of CVID patients with loss of memory B cells and expansion of CD21-low B cells which is particularly associated with autoimmunity and lymphadenopathy, and this patient probably falls into that category.

For now, the take-home message is that immune cytopenias, particularly ITP, commonly complicate CVID.

Question 46. TREC Newborn Screening for SCID

(see Primer of Clinical Immunobiology «Chapter—T-cell Immunodeficiency»)

46 (DAF*0939:2025). The frequency of TRECs (T-cell receptor excision circles) in newborn dried blood samples is emerging as a screening test for early diagnosis of SCID. The test itself is based on which physiological T-cell developmental process?

A. Negative selection
B. Formation of a T-cell receptor
C. Somatic hypermutation
D. Alpha-beta versus gamma-delta TCR decision

Answer B

Measurement of TRECs forms the basis for newborn screening for severe combined immunodeficiency (SCID). TRECs are a byproduct of T-cell receptor (TCR) gene rearrangement in the thymus. During formation of the TCRα chain by V-J recombination, the TCRδ locus, which is embedded within the α locus, is completely excised. The excised segment forms a circular DNA fragment, an 'excision circle', which remains episomal in the newly formed T cell. And because these excision circles are generated only during T-cell development, the number of TRECs in peripheral blood directly reflects the recent output of naïve T cells from the thymus.

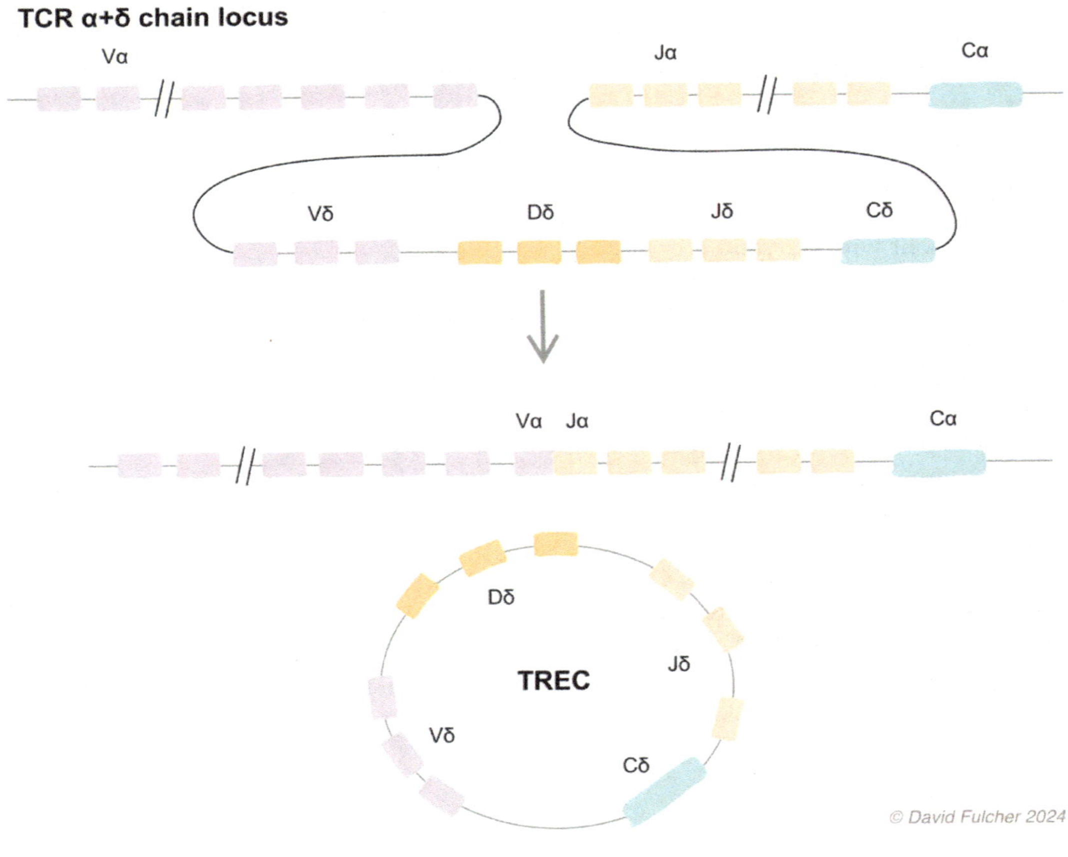

This principle underlies newborn screening for SCID. The infant's dried blood spot (Guthrie card) is tested for TREC content; if TRECs are low or undetectable, this suggests impaired thymic T-cell output and prompts the confirmatory step of directly quantifying circulating T cells by flow cytometry on a fresh blood sample. If confirmed, a diagnosis of SCID can be further explored (e.g. by genetics), and the child can receive definite treatment hopefully before getting infections.

Bottom line: TRECs are a surrogate marker of thymic activity, and their absence on newborn screening is a red flag for serious T-cell immunodeficiency.

Question 47. Antibiotic Allergy and Cephalosporin Cross-Reactivity

(see Primer of Clinical Immunobiology «Chapter—Non-Atopic Allergic Disease»)

47 (DAF*0940:2025). A 24-year-old man with cystic fibrosis presents with respiratory infection for antibiotic treatment. He gives a past history of penicillin allergy following a systemic anaphylaxis after penicillin G; skin tests were performed and were positive.

Which of the following would be the safest antibiotic choice?

A. ticarcillin
B. ceftriaxone
C. cephalothin
D. cephalexin

Answer B

This is a common scenario. Needless to say, based on history and skin testing, penicillins are essentially excluded for this patient, so we can forget about ticarcillin!

When it comes to cephalosporins, it's important to understand that the risk of cross-reactivity is primarily with the first-generation agents (e.g. cephalothin and cephalexin); there's little evidence of clinically significant cross-reactivity with any cephalosporin be-yond these first-generation drugs.

Cephalosporin	Drugs
1st generation	Cephalexin
	Cephalothin
	Cephazolin
2nd generation	Cefoxitin
	Cefuroxime
	Cefaclor
3rd generation	Ceftriaxone
	Cefotaxime
	Ceftazidime
4th generation	Cefepime

In other words, higher-generation cephalosporins, such as ceftriaxone, have minimal cross-reactivity and are generally considered safe in patients with IgE-mediated penicillin allergy. So in this case, ceftriaxone is the safest and most appropriate choice.

However please remember that acute penicillin desensitisation is an option for truly allergic patients, and can be administered over a day, thus allowing a first-choice (penicillin) antibiotic to be given if needed. Also remember that such desensitisation is only useful for the course that immediately follows, so the procedure will have to be repeated if any subsequent courses are given.

Question 48. UNC93B1 Mutations and Lupus Pathogenesis

(see Primer of Clinical Immunobiology «Chapter—Autoimmunity»)

48 (DAF*0941:2025). During the past year, four studies have reported ten mutations in UNC93B1, which encodes the Toll-like receptor (TLR) chaperone protein UNC93B1. All variants increased TLR7 and TLR8 signalling and caused systemic lupus erythematosus in young individuals. Such heightened signalling may explain the common association between lupus and:

A. the type I interferon signature
B. reduced complement levels
C. high CRP
D. hypergammaglobulinaemia

Answer A

A long way of asking, "what happens when TLRs (like TLR7/8) are tonically activated?!" We are told in the question that UNC93B1 mutations enhance signalling through TLR7 and TLR8, leading to excessive activation of these receptors. The corollary of this enhanced signalling is induction of type I interferons which drive the type I interferon signature.

Now a quick synopsis of the background to this question (although likely much more information than you will need!). The UNC93B1 gene product is a chaperone protein critical for the intracellular trafficking of toll-like receptors (TLRs), especially TLR3, TLR7, TLR8, and TLR9; it helps shuttle these receptors to endosomal compartments, where they recognise nucleic acids from pathogens (or self, in the case of autoimmune disease). Note that these four TLRs are not surface-expressed, but rather are found in endosomal compartments, perfectly positioned to detect internalised nucleic acids, whether from viruses or apoptotic debris.

Type I interferons are thought important in the pathogenesis of lupus in many patients, and likely explain the early onset of lupus in the UNC93B1 patients here. Even in adult lupus patients, at least half exhibit an interferon signature, meaning their cells show mRNA expression patterns that suggest exposure to type I interferons.

What about the other options?

- Reduced complement is a frequent finding in lupus, usually driven by immune complex formation, but these autoantigen/autoantibody complexes are driven by the adaptive, rather than the innate, immune system.
- CRP is typically low or at most modestly elevated in lupus, unlike systemic inflammatory disorders or bacterial infections.
- Hypergammaglobulinaemia is not uncommon in lupus due to B cell hyperactivity, but not directly related to TLR7/8 signalling; like immune complex disease, B-cell activation is part of the adaptive, rather than the innate, immune system.

So to summarise, UNC93B1 mutations enhance TLR7/8 activity, leading to increased type I interferon production, creating an internal 'milieu' conducive to development of lupus in these patients.

Question 49. SCID Mutations and Key Molecular Defects

(see Primer of Clinical Immunobiology «Chapter—T-cell Immunodeficiency»)

49 (DAF*0942:2025). Absence of circulating T cells might result from mutations in all the following EXCEPT:

A. RAG-1
B. CD40 ligand
C. IL-7 receptor alpha chain
D. Jak3

Answer B

Genetic absence of circulating T cells basically means severe combined immunodeficiency (SCID), and so the question is asking what mutations can cause it. Don't think that this question is only relevant to paediatricians; those with a handle on immunobiology should work through what each protein does, and thus whether its absence would affect T cell generation.

Let's walk through the options:

- RAG1 or RAG2 deficiency: These proteins are essential for V-D-J recombination. Without it, T and B cells cannot generate functional receptors, so both lineages will be absent. Result: T^-B^- SCID (see Question 25).
- CD40 ligand deficiency is the only defect that does not cause SCID, but rather X-linked hyper-IgM syndrome. CD40L is important for T cell : B cell collaboration, particularly class-switching, but impairs neither T nor B cell development.
- IL-7 receptor α chain deficiency: The IL-7 receptor is composed of IL-7Rα and the common gamma chain (γc) and is essential for T-cell development, so a deficiency in IL-7Rα (autosomal recessive) leads to T^-B^+ SCID.
- JAK3 deficiency: JAK3 is a signalling molecule associated with the γc, and is required for signal transduction from multiple cytokine receptors. Like γc deficiency, JAK3 mutations cause T^-B^+ SCID, but inheritance is autosomal recessive.

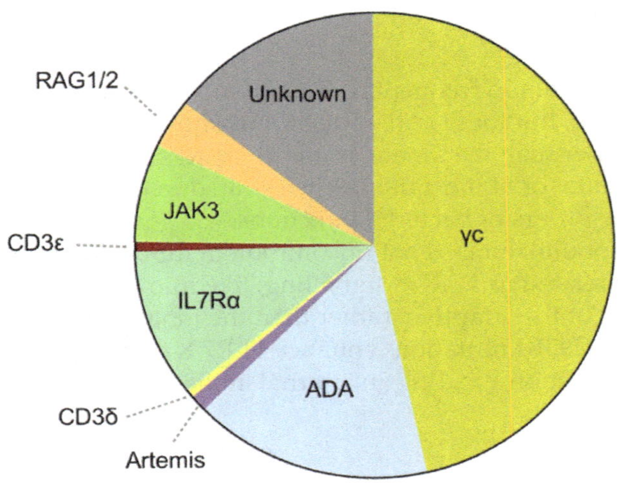

Note in passing: gamma common chain (IL2Rγ or γc) deficiency is the most common form of SCID, and is X-linked. γc is a component of the receptors of multiple cytokines

(IL-2, IL-4, IL-7, IL-9, IL-15, IL-21); its deficiency leads to T^-B^+ SCID, mostly because it affects IL-7R signalling which is needed for T-cell development. It was too obvious, so I didn't put it in as an option!

Question 50. Depemokimab and Eosinophilic Asthma

(see Primer of Clinical Immunobiology «Chapter—Pathophysiology of AllergicDisease»)

50 (DAF*0943:2025). Depemokimab is a long-acting monoclonal antibody to IL-5. As such, it is a potential therapeutic in the treatment of which of the following conditions?

A. Psoriatic spondyloarthropathy
B. Crohn's disease
C. Rheumatoid arthritis
D. Eosinophilic asthma

Answer D

You don't need to remember this difficult-to-pronounce name(!), just that it targets interleukin-5 (IL-5), and IL-5 is all about eosinophils!

IL-5 is produced predominantly by Th2 cells and is centrally involved in 'type 2' immune responses. It is the key cytokine driving eosinophil development, activation, migration and survival, so any disease with eosinophilic inflammation, e.g. eosinophilic (type 2) asthma, could potentially benefit therapeutically from IL-5 blockade. And it is well established that targeting IL-5 with mAbs like mepolizumab, benralizumab and now depemokimab has proven a highly effective strategy.

Regarding the other options:

- Psoriatic spondyloarthropathy is primarily Th17-driven.
- Crohn's disease is driven by Th1 and Th17 cells.
- Rheumatoid arthritis is predominantly driven by innate inflammatory cytokines (IL-6, TNF-α, IL-1) as well as autoantibodies.

Final Remarks

OK, that's it! I hope you did OK and proved to yourself that you have learned a lot about Clinical Immunobiology! To review anything that was covered in this Book and to continue your education journey, feel free to sign up to the key teaching resources available to support your learning:

- The Primer of Clinical Immunobiology;
- Vimeo 'immunology4physicians' lectures: immunology4physicians.vhx.tv
- Patreon 'immunology4physicians' revision exercises: patreon.com/immunology4physicians

Good luck!

David Fulcher

Made in the USA
Monee, IL
13 September 2026

63137525R00046